Rubina Patrícia Martins de Sousa

Conflict Management and Motivation of Hospitality Professionals

Rubina Patrícia Martins de Sousa

Conflict Management and Motivation of Hospitality Professionals

Human Resource Management

Imprint
Any brand names and product names mentioned in this book are subject to trademark, brand or patent protection and are trademarks or registered trademarks of their respective holders. The use of brand names, product names, common names, trade names, product descriptions etc. even without a particular marking in this work is in no way to be construed to mean that such names may be regarded as unrestricted in respect of trademark and brand protection legislation and could thus be used by anyone.

Cover image: www.ingimage.com

This book is a translation from the original published under ISBN 978-620-3-46820-5.

Publisher:
Sciencia Scripts
is a trademark of
Dodo Books Indian Ocean Ltd., member of the OmniScriptum S.R.L Publishing group
str. A.Russo 15, of. 61, Chisinau-2068, Republic of Moldova Europe
Printed at: see last page
ISBN: 978-620-4-09638-4

Acknowledgements

The completion of this research represents the conclusion of another important stage in my life. Thus, I could not fail to thank all the people who helped me on this journey and conveyed important values such as collaboration, cooperation, understanding and friendship.

To my supervisor, Professor Emília Fernandes for her availability, patience, support and valuable advice for the completion of this work.

To the hotel units that voluntarily offered to collect the sample data.

To my parents, João Amaro Sousa and Maria da Cruz Sousa and my sister, Joana for all the words of courage, help, affection, patience and understanding. Thank you for helping me build my future. Without you everything would be more difficult to accomplish.

To my friends Dr. Albertina Freitas and Dr. Jaime Freitas for all their words of motivation, support, help and generosity. Thank you for believing in me.

To all my friends for helping me overcome the difficult moments in the completion of this research. Thank you for giving me strength to complete this work.

To all my family and godparents, who in one way or another conveyed understanding, support and encouragement to carry out this research.

> *"Peace is not the absence of conflict*
> *but the presence of creative*
> *alternatives for responding to*
> *conflict."*
>
> *(Dorothy, Thompson)* [1]

---

[1] http://www.goodreads.com/author/quotes/75103.Dorothy_Thompson

Summary

Nowadays, conflict is an increasingly recurrent theme, both in professional and family life. The idea that the existence of conflict is synonymous with a lack of understanding, solidarity and collaboration still persists. However, conflict can also be constructive since it acts as a mechanism for correcting problems, increasing employee participation, innovation and organisational cohesion. Conflict will be constructive or destructive depending on people's motivation.

Motivation, in turn, is divided into two theories: the process theory and the content theory. McClelland's theory belongs to the content theory that tries to understand what motivates individuals to have certain behavior. McClelland divides his theory into three types of motives: the success motive, the affiliation motive and the power motive. While the affiliation motive is more linked to the affective bonds created within the organization, the success and power motives are more associated with standards of excellence and competitive and assertive activities.

This research aims to contribute with some further studies on how conflict management strategies are combined with motivational profiles. To this end, questionnaires were conducted to 349 individuals from the hospitality industry in the Autonomous Region of Madeira. The results suggest that: 1) the success motive correlates positively with the accommodation, commitment, avoidance and integration strategies; 2) the affiliation motive correlates negatively with the imposition strategy and positively with the accommodation, commitment, avoidance and integration strategies; 3) the power motive correlates positively with the imposition strategy and negatively with the accommodation and avoidance strategies.

The discussion of the results obtained from this research aims to contribute to understanding the relationship between conflict and motivation in the particular hospitality context and to point out human resource management measures that help to deal with conflict and employee motivation.

Keywords: Human Resources; Motivation; Conflict Management Strategies.

Abstract

Nowadays conflict is becoming a more and more recurrent subject, either in one's professional life, or at the personal level. The idea still remains that conflict raises from lack of understanding, solidarity and cooperation. Nevertheless conflict can also be positive as it may be the basis for problem solving, for the increase of staff participation, as well as for innovation and organizational cohesion. Conflict can be both positive and negative depending on people's motivation.

Motivation is based on two theories: the theory of process and the theory of contents. McCLelland's theory is included in the theory of contents which tries to understand what motivates individuals to adopt one particular behavior. McCLelland distinguishes between three types of motivation: achievement, affiliation and power. On one hand affiliation is connected to emotional bonds originated within the organization; on the other hand the motives of achievement and power are connected to excellence patterns and to competitive and assertive activities.

The current research aims at contributing with some additional studies on how the conflict management strategies coordinate with motivational profiles. To do so questionnaires were passed to 349 individuals working in the hotel business in Madeira Autonomous Region. Results suggest that: 1) the motive of achievement relates positively to strategies of accommodation, commitment, avoidance and inclusion; 2) the motive of affiliation relates negatively to strategies of imposition and positively to strategies of accommodation, commitment, avoidance and inclusion; 3) the motive of power relates positively to strategies of imposition and negatively to strategies of accommodation and avoidance.

The discussion on the conclusions of this research contributes to understand the relation between conflict and motivation, specifically in the context of the hotel business. It also points out human resources management policies that help dealing with conflict and motivation of the staff.

Key words: Human resources; motivation; conflict management strategies

Table of Contents:

List of Abbreviations

R.A.M. - Autonomous Region of Madeira

V.A.B. - Gross value added

Introduction

The theme chosen for the present work is "Conflict Management and Motivation of Hospitality Professionals". The choice of this theme is due to the fact that society is constantly changing and organizations increasingly seek the maximization of the skills of their professionals enriching their knowledge, allowing a better flexibility and rationalization of resources (Chiavenato, 1999). For this to happen, it requires people who know their employees, who are able to manage conflict effectively and promote the motivation of professionals within the organisation, contributing to organisational success.

According to McIntyre (2007), the increase in competitiveness requires the formation of heterogeneous teams capable of performing multiple functions in the most diverse organisational departments. As these heterogeneous teams are composed of people with different ideas and perspectives, conflict, and consequently its management, becomes inevitable (McIntyre, 2007). The area of conflict is actually as important as the areas of planning, communication, motivation and decision-making (Bilhim, 1996). While it may represent a constraint for the organisation, conflict can also become an advantage for it, since, when constructive, it allows for greater participation between groups, enables greater mutual understanding, generates greater consensus about agreements and decisions, and favours intragroup collaboration (Cunha, Rego, Cunha, & Cardoso 2007). In addition, it allows for the sharing of opinions, increases the likelihood of creative solutions, reduces social laziness and increases the quality of decision-making (Cunha et al., 2007). Thus, we can say that the presence of conflict is important to the extent that it increases individual, group or even organisational performance levels (Cunha et al., 2007).

On the other hand, in a competitive market, organizations should be increasingly concerned with the skills and characteristics of their employees so as to gain a competitive advantage over other organizations. One of these important characteristics is the employees' motivation, which reflects the desire to join efforts to achieve organizational goals, being these same efforts a necessary condition for the satisfaction of the individuals' needs (Ferreira, Neves, & Caetano, 2011). In addition, motives prove to be an important component for the organisation to the extent that "they occupy an important place in the list of the various psychological elements that make up the personality of individuals" (Winter referred to by Rego & Carvalho, 2002, p.17).

Over the last few years, the concept of conflict and motivation has been the subject of extensive research by some researchers who try to understand the combination of motivational variables with conflict strategies (Rego, 1995; Rego & Jesuíno, 2002). Thus, this study aims to explore the impact of imposition, accommodation, avoidance, commitment and integration strategies on the motives of success, affiliation and power, in the hotel context in Madeira. More specifically, this study aims to: a) characterize the conflicts present in Madeira's hospitality companies; b) characterize the motives present in the region's hospitality sector; c) evaluate the relationship between conflict and motivation in the region's hospitality professionals; d) show the importance and need for a good conflict management and motivation of professionals for organizational success.

The target population of the study is the companies in the hospitality industry, located in the Autonomous Region of Madeira (R.A.M.). For this purpose 27 companies were surveyed, which constitute our sample. In order to carry out this study, a quantitative methodology was used, using a questionnaire addressed to 349 hospitality professionals in the Madeira Autonomous Region. In this sense, and in order to achieve the research objective, we used Jesuíno's "conflict management strategies" questionnaire (2003 adapted from Thomas, 1976) developed with the objective of assessing the most predominant conflict strategy among hospitality professionals in the Madeira Autonomous Region.A.A.M. We also used the "questionnaire measuring the motives of success, affiliation and power" of Rego and Carvalho (2001, 2002) with the purpose of assessing the most predominant type of motive in hospitality professionals in the R.A.M.

As regards the structure of the study, it is divided into several parts. In the first part, which deals with the theme of conflict, it is considered important to define conflict, understand how conflict is created in organizations, understand its advantages and disadvantages, and strategies for its resolution.

In the second part, concerning motivation, it becomes important to define the concept of motivation, understand which theories of content, understanding the specifics of the theory of David McClelland.

In a third part, the characterization of the hotel industry is also presented, including the characterization of the hotel industry in the Autonomous Region of Madeira and the professionals working in it. Subsequently, in the methodology, it is considered relevant to present the sample, the data collection instrument and explain how the data will be collected. After the statistical treatment of the data, the results and the discussion of the work are presented, associating the theoretical contents to the practical work.

Finally, the main conclusions of the paper, the limitations of the study and suggestions for future research are   presented.

Chapter I: Conflict in organizations

1.1 Concept of Conflict

In recent times, the concept of conflict has been the subject of several studies by the most varied and important researchers (Chrispino, 2007; Caetano & Vala, 2000; Chiavenato, 1999). Thus, and in the context of this work, it is important to define the concept of conflict, its characteristics and the impact it has on the organization.

McIntyre (2007, p.296) states that organisations are often embarrassed to admit the existence of conflict, as they associate the term conflict with "a failure on their part, a relative loss of control of their employees, being an indicator that the organisation is not functioning well". However Thomas (1992, p.265) advocates conflict as "the process that begins when one party perceives that the other is frustrated, or is about to be frustrated ". The classical theories of Fayol, Gulick and Urwick, Taylor and Weber (referred to by Rahim, 2001) argue that the existence of conflict is a harmful factor for the effectiveness of the organization, so it should be reduced. According to these same authors, through harmony, cooperation and conflict elimination, it is possible to achieve greater organizational effectiveness.

However, as conflict is a complex phenomenon, there is no simple definition for this topic. Thus, Pruitt and Rubin (1986, p.4) state that conflict exists due to the divergence of interests or the conviction that it is impossible to simultaneously achieve these same interests or aspirations. For his part, Goel (2012) states that the conflict arises when there is an incompatibility of interests originating an opposition of objectives, motivations and real or inferred actions by the two parties in conflict. For Férnandez and Ríos (referred to by Cunha, 2008), conflict arises between two or more individuals or groups who, when interacting, show an incompatibility of behaviours (internal or external), causing one of the parties to exert power over the other. However, Kolb and Putnam (1992, p. 312) state that we are facing a conflict when "there are real or perceived disparities that appear in certain organisational contexts giving rise to emotional states as a consequence".

For Freire (1942) the conflict at work stems from strong events and is a predictable phenomenon in that it requires prior reasoning of one of the parties in conflict. Regarding the conflict, Chiavenato (1999) states that it occurs between line managers and human resource specialists, when they are responsible for decision-making about people management, thus existing a dispute about the individual who holds the power to make decisions about people or due to divergent orientations.

To characterize the conflict, Robbins (2002) defines three important views: the rational view, the human relations view and the interactionist view. According to the same author, the first view is the oldest and argues that the conflict should be avoided, since it is seen as something harmful to the organization. The second view, as defined by Robbins (2002), on human relationships, argues that in any group, the conflict is something natural and inevitable, not necessarily harmful to the organization, and may have a positive effect on group performance. Poudy (referred to by Rocha, 2010) shares the same opinion when he defines conflict as something natural and inevitable, disassociating the term conflict from the mismanagement of managers and administrators. Finally, the interactionist view is defined by Robbins (2002) as one that treats conflict as a contribution to change and innovation, i.e., if leaders know how to manage conflict, the group remains viable, innovative and critical, removing the need to avoid conflict. Simpson (1998) summarises these three visions by stating that if in the past conflict was synonymous with violence and reprimand, today it is seen as something challenging and motivating, and in this sense, an opportunity for growth and personal development.

For Moscovici (2001) and Muchinsky (2004), conflicts that help groups and organizations to achieve and realize their intended goals are considered functional, and conversely, those that hinder the achievement of goals are considered dysfunctional conflicts.

1.2 Advantages and Disadvantages of Conflict

The advantages or disadvantages of conflict differ according to how it is managed. Whether the conflict is positive or negative, it will always affect two variables: the individual and the organisation as a whole. However, conflict can lead positively to the organization and negatively to the individual or vice versa.

For Baron (referred to by Cavalcanti, 2006), destructive conflict is that which hinders communication, generates low coordination, creates biases, causes autocratic leadership change and reduces the search for other perspectives. For his part,

Chiavenato (1998) states that conflict causes frustration, hostility and anxiety, creates pressure for the conformity of individuals, disperses energies, creates blockages and refusals in cooperation and, finally, generates misperceptions. Additionally, and according to Rahim (2001), conflict is harmful to the organisation when: a) it generates *stress* and dissatisfaction for the organisation; b) it creates a climate of distrust and suspicion; c) it is harmful to relationships; d) work performance is negatively affected; e) organisational commitment and loyalty may be undermined. Other negative effects associated with the notion of conflict can also be seen in Table 1:

Table 1 - Destructive effects of conflict

Negative effects
It destroys the morale of groups and organizations.
It induces opponents to make hostile attributions of others.
It polarizes groups and individuals, deepens and widens differences.
It creates irresponsible and retaliatory behavior.
It turns the organization into a "political arena".
It creates a paranoid atmosphere.
It ruins some people's careers.
Increases absenteeism and *turnover.*
Reduces organizational commitment.
Suscribes the destruction of the group.
It diverts energy from more important tasks.
It generates an atmosphere of zero-sum orientation.
Leaders move from a participative to directive leadership style

Source: Adapted from Cunha et al (2007, p.536).

Several studies on the impact of conflict on group functioning have shown that conflict is a harmful factor for the work team, since it reduces group unity, destroys communication networks, and decreases productivity (Deutsch, Guetzkom & Gyr, Schwenk & Cosier referred to by Dimas, Lourenço, & Miguez, 2007). However, according to Cunha et al (2007) the level of conflict at an optimal level allows increasing creativity, quality of decision-making and performance. While for Baron (mentioned by Cavalcanti, 2006) the positive consequences of conflict are related to interest and new ideas, improved performance, and increased interest in key problems, for Chiavenato (1998), constructive conflict is the one that arouses interest and curiosity, increases group unity, increases motivation for the task, draws attention to problems, and, finally, analyzes and reduces power differences. It should also be noted that, for Rahim (2001), constructive conflict is one that: a) stimulates innovation, creativity and growth; b) improves decision-making within organizations; c) finds alternatives to a problem; d) leads to synergistic solutions to common problems, e) improves individual and group performance; f) creates new approaches for individuals and groups; g) can force individuals and groups to articulate and clarify their positions. According to Tjosvold (2008), positive conflict is that which involves cooperation and is closely related to the conviction of individuals to achieve certain objectives and goals. According to the same author, positive conflict involves debates that allow for an open exchange of opinions and points of view that lead to a better understanding of the other, enabling the exchange of ideas and allowing for mutual benefit. According to Robbins (2002), functional conflicts translate into low levels of process conflicts (interpersonal relationships) and low to moderate levels of task conflicts (related to the content and objectives of the work), since they help in the debate of ideas, increasing organizational performance as a result. According to Tjosvold (2008), constructive conflict in the short and long term is that which helps to understand certain issues, helps to find quality solutions, and strengthens relationships.

However, the motivation of people is what determines the choice for a constructive or destructive conflict, being fundamental the presence of the manager or manager to facilitate conflict management (McIntyre, 2007).

In short, conflict has both positive and negative consequences. However, we should enhance the conflict, reducing the negative effects and strengthening the positive ones (Rahim, 2001). Thus, if we manage conflict in order to strengthen its positive effects, we are increasing organisational commitment, increasing innovation and solving individual and group problems in a creative way, avoiding the destruction of groups and the diversion of energies that would be important for the performance of certain tasks.

1.3 Conflict Management Strategies

In general, conflict resolution involves obtaining a "discernible result", ending conflict conduct, and distributing values and resources in an acceptable way (Bercovitch mentioned by Cunha, 2001, p.47). By this the author means that, in general, the solution to the problem of conflict is through the removal of the conflict situation and the acceptable distribution of benefits arising from this situation.

In conflict resolution strategies, negotiation stands out as a way to deal with conflict. Morley and Stephenson (referred to by Cunha, 2001, p. 51) define negotiation as "the direct or indirect verbal (or non-verbal) communication through which the parties to a conflict of interest discuss, without recourse to arbitration or other judicial processes, the form of any joint action they may take to manage the dispute between them". In this sense, negotiation is a form of conflict resolution between two or more individuals, in which there is an adequacy of wills between the various parties in order to find a situation capable of partially satisfying the interests of both sides (Serrano referred by Cunha 2001).

The main objectives of negotiation are: the achievement of positive results, the influence on the balance of power, the development of a constructive climate and the achievement of a flexible dynamic (Cunha, 2003). The first objective concerning the achievement of positive results is governed by the principle of problem solving which is based on the "firm/flexible" position, referred by Pruitt (mentioned by Cunha, 2003, p.2). This position advocates firmness in defining elementary interests and flexibility in the means to achieve those same interests. However, when faced with the "concession/hardness" position, the negotiator should use tenacity and firmness without ever closing to concessions. Another objective of negotiation is the influence on power equality, which states that for a negotiation activity to be constructive it is necessary to have power equality, that is, even if there is a propensity for domination, the purpose is to achieve more advantages and actions (initiatives) than the adversary, besides which it is important that both parties in conflict acknowledge their dependence on each other (Cunha, 2003). Another objective of the negotiation is the development of a constructive climate, being necessary for this a greater capacity of communication between those involved. In this sense, the negotiator must keep the balance, avoiding excesses, whether of joviality or hostility, because this balance will confer credibility to the negotiation and a greater possibility of acceptance of interdependence (Cunha, 2003,p.2). Finally, we have the establishment of a flexible dynamic as a negotiating objective, in which Pruit's principle (referred to by Cunha, 2003,p.2) "firmness/flexibility" is also used, but without resorting to concessions, i.e., "firmness" is associated with the means and "flexibility" with the objectives or self-interests without using concessions, examining the association of collaboration with competition and the sharing of interests, promoting organizational integration.

According to Rahim (2001), negotiation does not modify the processes and structures of the organisation, that is, it does not change the functioning of the organisations. However, and according to Cunha (2003), negotiation cannot be considered a valid solution for all problems, since not everything is negotiable (for example: conflicts of beliefs and values). Additionally, and as seen above, the conflict should not always be avoided, since it can bring positive effects, such as creativity and innovation essential to the success of the organization.

Still with regard to negotiation, we may consider the dimensions of assertiveness and cooperation. These two dimensions show the motivational orientations of each individual when facing a conflict situation (Marques & Cunha, 2004). According to Thomas (1992) assertiveness occurs when there is a concern to satisfy one's own interests, while cooperation tries to satisfy the interests of others. The combination of these two dimensions leads to the five conflict management strategies: accommodation, competition, collaboration, avoidance and commitment (Thomas 1992). Figure 1 represents the combination of the two variables which gave rise to the strategies described by Thomas (1992):

Figure 1 - Combining cooperation with assertiveness

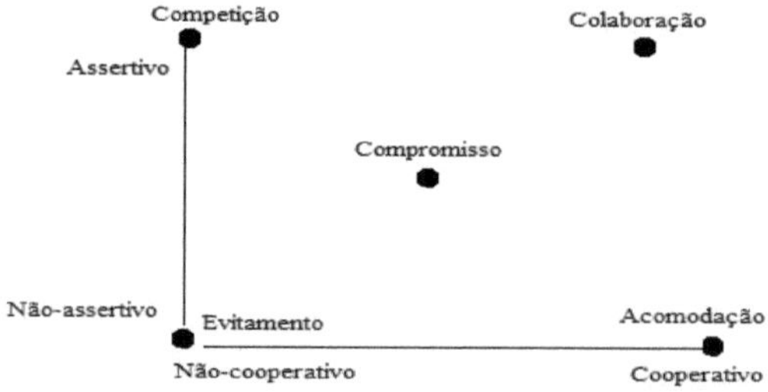

Source: Thomas (1992,p.266)

With regard to the accommodation strategy, and according to Karakus and Savas (2012), we found that, when using the accommodative strategy, the individual shows a low concern for satisfying his/her own interests and a high concern for satisfying the interests of others. According to Karakus and Savas (2012), in the accommodative style, the individual tries to decrease disparities and marked similarities, showing a high concern for the other, accepting and reconciling the desires of other individuals. The accommodating individual reacts differently depending on the type of conflict he/she faces. Thus, according to Ferreira, Neves, and Caetano (2011), if we are considering the conflict of objectives, for example, accommodation translates into finding a solution that meets the other party's objectives, neglecting one's own objectives. According to the same authors, in the case of an interpretation conflict, the accommodated individual shares the others' opinion, even without agreeing with it. If we are facing a normative conflict, the individual ignores the perceived infraction, allowing the other party to persist with the committed infraction (Ferreira, Neves, & Caetano, 2011). According to Antonioni (1998), the accommodative style is characterised by a loss-gain orientation and is considered a self-sacrifice style. The accommodative strategy proves to be more appropriate when: a) we believe that the other party is right; b) the issue is more important to the other party; c) we give in now and expect to receive something in return in the future; d) we are in an inferior position; e) it is important to preserve the relationship (Rahim 2001, p. 83).

In relation to the competitive or dominative strategy, the individual who uses this strategy is the one who shows high concern for him/herself and low concern for others (Munduate, Ganaza, Alcaide, 1993). In this strategy, the individual tries to improve his results neglecting the opponent's results (Moberg, 2001). According to McIntyre (2007), in the case where the individual is in a superior position, he/she will take advantage of his/her power situation to force subordinates to obey and fulfil his/her wishes. Thus, this strategy is associated with a win-lose relationship, since the individual who tends to be competitive uses all means to achieve his or her goal, often ignoring the desires and expectations of his or her opponents (Munduate, Ganaza, Alcaide, 1993). Additionally, Karakus and Savas (2012) state that, in this win-lose situation, the subject imposes a certain behaviour to the opponent to win the other's position, and may use threats, illusions, persuasive arguments and positional promises. According to Dimas, Lourenço and Miguez (2005), and as this is a win-lose strategy, it does not allow for a mutual solution and results only in the short term. According to Rahim (2001, p.83), the competitive style is more appropriate when: a) the issues are important to the work group; b) the other group makes an unfavourable decision that harms my group; c) a supervisor covers routine issues or when a quick decision is required; d) the supervisor deals with subordinates who are very assertive and have little experience in making technical decisions; e) an implemented training course is not appreciated by all. In turn, McIntyre (2007) states that the competitive strategy is more used when it becomes imperative to defend one's rights and/or in cases where the individual is secure in his/her position.

According to Karakus and Savas (2012), the avoidance strategy results from the low concern with satisfying one's own interests and low concern with satisfying the interests of others. According to these same authors, the individual with a tendency towards this strategy is the one who, in a conflict situation, tries to go unnoticed or avoid the situation. Dimas, Lourenço and Miguez (2005) take a similar position, stating that the avoidance strategy arises when the individual runs away from the conflict situation or denies the existence of any problem. According to these same authors, the use of avoidance as a way to overcome conflict is often the postponement of problem solving or the cessation of a threatening situation. According to Rahim (2001), an individual who uses this style of conflict is the one who assumes an attitude of contempt before the problems or the parties in conflict, does not assume responsibilities, transferring them to other people and denies the existence of conflict in public. Still on this conflict strategy, Antonioni (1998) mentions that the avoidance strategy is characterised by a lose-lose orientation, since the subject does not tell his/her needs to the other party in conflict, ending up being dissatisfied. McIntyre (2007) complements this theory by stating that this loss-lose relationship arises when there is an impossibility of mutual achievement of what is intended and when the reasons for the conflict remain unknown. For Rahim (2001, p.83) this strategy is more appropriate when: a) it pays to confront the other party, since the negative effect of confrontation outweighs the positive effect of conflict resolution; b) the issue is trivial; c) a period of reflection is necessary. In turn, McIntyre (2007, p.299) states that the avoidance strategy is used when: a) the issue is unimportant; b) there is no possibility of winning; c) it is necessary to spend more time to gather information or a disagreement may be dangerous.

In turn, the compromise or mutual concession style is characterised by an intermediate concern with oneself and others, i.e. a mix of cooperation and assertiveness (Karukus & Savas, 2012). According to Cunha et al (2007), this strategy represents an attempt to satisfy the interests of both parties in conflict. Already Munduate, Ganaza and Alcaide (1993) argue that both parties involved in conflict give in something to achieve an acceptable situation for both sides, this is achieved through an exchange of compromise or search for a middle position. According to Rahim (2001, p.30) this strategy arises when "the compromising party gives up more than the dominating party, but less than the accommodating party". For Antonioni (1998) despite the attempt at mutual solution, and contrary to the integrative style, the compromise style prevents the satisfaction of all the needs of both parties in conflict. McIntyre (2007) complements by stating that the compromise strategy translates into a win-lose situation, since it becomes impossible to fully satisfy needs for both sides. This strategy is most appropriate when (Rahim, 2001, p.83 and p.84): a) the objectives of both parties are exclusive; b) both parties are equally strong and cannot find a solution through negotiation; d) individuals need a momentary solution to a complicated problem; e) other strategies have proved ineffective in solving the problem; f) it can be used to end the conflict more quickly (avoiding prolonged conflict).

Finally, the collaborative or integrative style is more cooperative and assertive, i.e. it shows a high concern for oneself and others (Thomas, 1992). This conflict resolution strategy is characterised by open communication in which, through information sharing, a desirable situation is achieved for both parties in conflict (Dimas, Lourenço, & Miguez, 2005). In this way, individuals constructively explore different problems and seek solutions to reach a satisfaction of mutual interests. This attempt at mutual satisfaction may result (Cunha et al., 2007, p.526) in a "win-win" solution that allows for the total achievement of the defined objectives; the achievement of an idea or conclusion convenient for both parties in conflict or the transmission of a set of expectations and interpretations, trying to reach an agreement on what is or is not acceptable in a given situation. Individuals who tend to be collaborative acknowledge different opinions and try to learn something from other individuals' opinions in order to reach common solutions (Christopher Moore's referred to by Chinyowa, 2013). Thus, collaboration seems to be the most favorable strategy for individuals since it proves to be appropriate when: (a) it is necessary to deal with complex problems; (b) the problem cannot be solved only by one party (being necessary the synthesis of ideas to reach a better solution to the problem); (c) it is used skills, information and other resources to define or redefine a problem; d) it is important to reach an effective alternative solution and/or when the commitment of both parties to the conflict is necessary to reach an effective implementation of the solution (requiring the expenditure of time); e) it deals with strategic issues relating to objectives and policies of the organization that involve long-term planning (Rahim, 2001, p.81). In this sense, the integrative style seems the best style for individuals to use. However, there are certain situations in which this is not the best style to use. These are: urgent decision making or when there is disinterest by one of the parties in relation to the results or when the parties of the conflict are not able to solve the conflict (Rahim, 1992 referred by Antonioni, 1998). Thomas (1992) studied the long-term benefits of the collaboration strategy, concluding that collaboration is a desirable situation for individuals and organizations.

The main objective in conflict management is to find a satisfactory solution for both parties involved in the conflict, minimizing its negative effects (Chinyowa, 2013). Tjosvold (2008) takes a similar position stating that conflict management is a much more effective joint activity when both sides believe in the mutual attempt of conflict resolution in order to benefit all individuals.

With regard to the five previously mentioned conflict strategies, Rahim (2001) considers that when dealing with strategic or complex problems, the integrative or collaborative strategy is the most appropriate. However, when we are talking about tactical and routine (day-to-day) conflicts, the strategies of accommodation, compromise, avoidance, and competition may be more effective when applied appropriately to each situation. About the benefits of the integrative strategy, McIntyre (2007) states that, if this strategy is chosen, we may reduce psychological distances and increase communication, thus favouring a climate of trust conducive to a more creative and lasting conflict resolution. According to Tjosvold (2008), the studies previously conducted on conflict durability state that the use of cooperation as a means of problem solving brings about both short- and long-term beneficial consequences.

However, conflict management strategies prove to be more or less appropriate depending on the situations faced, and there is no single and/or most appropriate management strategy for all situations (Thomas; Rahim & Bonoma referred to by McIntyre, 2007).

It should also be noted that, of the five conflict management strategies studied, the most commonly used is undoubtedly the collaborative strategy and the least used is the avoidance strategy (Rahim & Buntzman mentioned by Dimas, Lourenço, & Miguez 2005). On the contrary, and according to McIntyre (2007), in a situation in which individuals feel uncomfortable to talk, they are more likely to use the avoidance or accommodation strategy, which is more harmful to the organisation.

To sum up, for conflict management to be effective there has to be negotiation and an ideal negotiation process always implies compromise. This means that we have to give in to the other party something that we often consider important, only then will a mutual agreement be reached. Therefore, the ideal is to make concessions on both sides, so that it becomes a fair and satisfactory conflict resolution for both sides of the conflict and, above all, productive for the organization.

Chapter II: Motivation of professionals

2.1 Concept of Motivation

Whatever the activity, individuals are dependent on personal motivation to support certain efforts (Bruce & Pepitone, 2002). Motivation is an integral part of the human being. Since motivation is a complex phenomenon, there are several definitions for the same theme. Motivation, according to Koenes (1996, p. 191), implies "an emotional state generated in a person as a consequence of the influence exerted by certain motives on his or her behaviour". Whereas for Robinson (2000), and in organizational context, motivation occurs when people, in a free and enthusiastic way, do what is required of them by their profession, organization or business. Also, according to Rocha (2010, p.104), motivation is: "a complex phenomenon, not purely individual, but that results from the interaction between individuals and situational variables".

Thus, according to the definition of Rocha (2010) motivation does not derive only from the individual, but also from a set of people (employer and colleagues, for example) who encourage them to achieve better results, reaching a higher level of personal satisfaction. Motivation can be interpreted differently and taking into account the context and situation of the organisation. About organisational motivation, Houston (referred to by Rocha, 2010) argues for the existence of motivational differences between the public and private sectors. This study concluded that civil servants feel less motivated to work than private sector employees, since those in the private sector have more financial incentives. On the other hand, Lacerda and Abbad (2003,p.82) when studying training and its motivational variables defined motivation for training as "the direction, effort, intensity and persistence with which trainees are interested in learning-oriented activities before, during and after training".

Motivation is also directly linked to organizational goals (Uroševic & Milijić, 2012). According to the same authors, organizational motivation is considered efficient if it meets personal needs and organizational objectives. Thus, more motivated employees know how to act and how to achieve the predefined objectives more quickly and effectively, benefiting the employee and the organization as a whole, since their effort and work are rewarded (Uroševic & Milijić, 2012,p.175). Also for Rocha (2010) the motivation of a given individual arises when their behavior is directed to the achievement of a certain goal. Although motivation is not easy to identify, an individual shows to be motivated, or not, for a certain goal depending on the behavior observed at a given time (Neves, 1998). According to Bruce and Pepitone (2002), people feel more motivated to do what arouses their interest, which may result in successful or not successful behaviors, being the successful behaviors considered satisfactory in the daily lives of people.

In the area of human resource management, it is essential to refer to motivation from the management perspective. However, we recall that management does not act directly on people's motivation; it can only interfere in what motivates them (Bruce & Pepitone, 2002). According to the same authors, the manager or the management policies may be primarily responsible for positively or negatively influencing what motivates employees, increasing the performance and efficiency of their functions. Uroševic and Milijić (2012) take a similar position by stating that it is necessary for management to perform a set of complex activities inherent to increasing motivation and organizational satisfaction. A good perception of motivation leads to an increase in efficiency and creativity, an improvement in lifestyle, and also increases the competitive edge and organizational success (Uroševic & Milijić, 2012). According to these same authors, in order to increase personal motivation, we need to define our limitations, make safe choices, adapt our work to our private life, define new challenges, set clear goals, improve and develop new skills, and, finally, collaborate and help others.

According to Uroševic and Milijić (2012) there are several theories of motivation that answer basic questions, such as, what motivates human activity and how the process of motivation takes place. Therefore, there are two major groups of theories of motivation: content theories and process theories. Content motivation is based on the causes of motivation, while process motivation is based on the need to seek to understand how the process of motivation arose (Uroševic & Milijić, 2012).

2.2 Content Theories:

The content theory of motivation explains what motivates people's behaviour, i.e., this theory aims to understand which factors lead the individual to behave in a certain way (Rego, 2000). Cunha et al (2007, p. 155) share the same opinion stating that content theories try to explain motivation through its causers, answering the question:

what motivates people? From this question we assume the existence of certain inner needs that need to be satisfied and which translate into the source of energy of human behaviour (Camilleri, 2007).

There are several theories of motivation, the content theories (e.g.: Maslow's hierarchy, Alderfer's ERG theory and Herzberg's theory) and the process theories (e.g.: Adams' theory, Vroom's theory, Locke's theory of goals, among others) however, and taking into account our study hypotheses, we delved into the content theories and David McClelland's theory.

2.2.1 Maslow's Hierarchy of Needs

In 1943 Maslow (referred to by Chiavenato, 1987) formulated the hierarchy of needs that influence human behavior, translated through the image of a pyramid whose base is composed of the most basic needs and ends in the more complex ones.

Figure 2 - Maslow's pyramid of needs

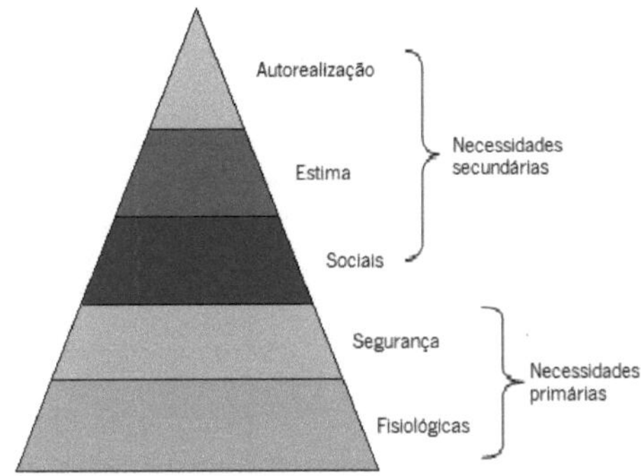

Source: Chiavenato (1987)

The primary needs are the physiological and safety needs, while the secondary needs refer to the social needs, esteem and self-actualization (Chiavenato, 1987). According to Chiavenato (1987) and, starting from the most basic needs to the highest, physiological needs refer to survival needs, for example: air, food, rest, shelter, etc.; safety needs refer to protection against hostilities or deprivation; social needs refer to friendship and acceptance by groups; esteem needs refer to the needs for consideration, recognition, love and self-respect; and, finally, self-actualization needs refer to personal fulfillment and full use of their individual skills. According to Ferreira, Neves, and Caetano (2011), individual differences are less visible in primary needs than in secondary needs, the former being more urgent than the latter. This is explained by the fact that primary needs are related to survival needs, which are common to every citizen. Maslow's theory is summarized in figure 2.

As seen previously, the hierarchy of needs assumes an order in which it begins in the physiological needs and ends in the self-actualization needs. According to Ferreira, Neves, and Caetano (2011), needs motivate the individual while they are not satisfied, i.e. once the primary needs are satisfied, the secondary needs have greater power to motivate the individual. Additionally Maslow (referred to by Noltemeyer, Bush, Patton & Bergen, 2012, p.1863) calls this hierarchical order "relative prepotency", in which after the initial satisfaction of the basic needs other higher level needs immediately arise that also need to be satisfied. However, Maslow (referred to by Noltemeyer, Bush, Patton & Bergen, 2012) states that, although at a given moment one need dominates the other, the individual may be motivated for several needs at the same time. Additionally, according to these same authors, an individual who is at the top level of the pyramid (needs for growth and self-actualization) may later return to the initial level concerning the satisfaction of basic needs. For example: in a situation of family unemployment, we are forced to temporarily sacrifice the needs for growth and self-fulfillment and, on the other hand, to satisfy the most urgent basic needs necessary for human survival (Noltemeyer, Bush, Patton & Bergen, 2012).

Also according to Ferreira et al (2011), in Maslow's theory, the needs for achievement and esteem encourage individuals to improve their performance in the workplace. This means that individuals with needs for recognition, appreciation and personal/professional growth have a better *performance* and, consequently, greater success in the workplace. Thus, for an individual to reach the ideal level of functioning of Maslow's needs, it is necessary for the basic and secondary needs to be completely satisfied (Noltemeyer, Bush, Patton, & Bergen, 2012).

2.2.2 Alderfer's ERG theory

Alderfer's theory (referred to by Pérez-Ramos, 1990) comprises three major groups of needs: existence needs, relationship needs and growth needs. In comparison to Maslow's theory of needs, the existence needs (physiological well-being and extra rewards offered) are related to physiological and safety needs; in turn, the relationship needs (creation of favourable interpersonal relationships associated with the desire for socialisation and interaction) are associated with social and esteem needs; and, finally, the growth needs (personal fulfilment) are related to self-actualisation needs (Cunha et al., 2007).

Additionally, while the satisfaction of each level of Maslow's pyramid works in a progressive and increasing manner, Alderfer's theory of needs (referred to by Pérez-Ramos, 1990) works in the opposite direction, i.e., it works in a regressive and decreasing manner (frustration-regression). In this way, the non-satisfaction of higher-level needs may constitute a blockage or a frustration, leading the individual to activate the lower needs already previously satisfied (Schermerhorn, Hunt, & Osborn, 1999).

Still on this frustration-regression dichotomy, Alderfer (mentioned by Stoner & Freeman, 1999) admits a certain variation in the directions of the hierarchy of needs, i.e., depending on the situation, individuals ascend or descend within the same hierarchy of needs. According to Pérez-Ramos (1990), Alderfer's theory is more flexible than Maslow's theory, since it allows for the satisfaction of more than one need simultaneously.

2.2.3 Herzberg's Two Factor Theory

The main objective in Herzbeg's study (referred to by Ferreira et al, 2011) is to try to understand which factors generate satisfaction or dissatisfaction in the workplace, dividing them into intrinsic and extrinsic factors.

According to the author, the main reasons for satisfaction at the workplace are associated with intrinsic factors, such as responsibility, growth, development, achievement and recognition, and the nature of the work. For Herzberg (referred to by Ferreira et al., 2011), these factors are the ones that most motivate individuals, since they are related to individual performance and the upper levels of Maslow's pyramid (social needs, esteem, and self-actualisation), as shown in Table 2.

Table 2 - Comparison of Maslow's and Herzberg's theory

Motivating factors	Hygienic factors
Feeling of accomplishment	Relationship with the boss
Recognition	Relationship with colleagues
Varied and challenging work	Technical supervision
Personal development	Working conditions

Source: Cunha et al (2007).

On the other hand, job dissatisfaction factors are associated with extrinsic or hygienic factors related to the work environment. The designation of "hygienic" factors is due to the notion of mental health in the area of psychiatry, considering these factors as necessary but not sufficient, since they prevent "negative attitudes but do not cause positive attitudes" (Cunha et al, 2007, p.161). Herzberg (referred to by Ferreira et al., 2011) also associates extrinsic factors with management policies, leadership styles, the relationship with the hierarchical superior, working conditions, salaries and interpersonal relationships, among others. It should also be noted that Herzberg (referred to by Miner, 2005, p.65) associates money with a hygiene factor stating that "because of its ubiquitous nature, salary usually appears as a motivator, as does hygiene. Although hygiene is a primary factor, it also often takes on some of the properties of a motivating factor with dynamics similar to those of recognition and achievement. ”

2.2.4 David McClelland's Theory of Motivation

David McClelland was a distinguished professor and researcher of psychology at Boston University, having also served as professor of psychology at Harvard University. This author was responsible for the development of motivation theory that distinguishes three important motives: the success motive, affiliation and power.

Success-driven individuals desire excellence, are known for their initiative and entrepreneurial spirit (Rego & Leite, 2003). These people are typically restless in their workplace, like to take moderate risks and seek *feedback* from others in order to increase performance levels (Rego & Leite, 2003).

With regard to goal setting, Rego and Jesuíno (2002) state that individuals who tend to be motivated by the success motive want to achieve high but realistic goals. According to these authors, in addition to the taste for high goals, these individuals respond positively to competition and like to take personal responsibility in the activities they perform, thus finding solutions to a wide variety of problems.

Regarding the execution of tasks and, according to Rego (1995), individuals oriented towards this motive prefer moderately difficult tasks, i.e., individuals tend to be perfectionists who seek to "always do better". McClelland (1987) adds, stating that individuals who tend to be motivated for success prefer moderate risk, denoting a better performance of equally moderate risk tasks.

The study of Rego and Carvalho (2002) conducted with higher education students justifies the high performance in individuals who tend to be motivated to succeed by the fact that academic practice leads students to receive *feedback* about their performance, making them closely responsible for the results acquired.

When it comes to small business managers, they seem to be more motivated for success than power, since in small businesses, improving and operationalizing the business in terms of efficiency/cost is more urgent than the ability to influence people (McClelland, 1965).

In conclusion, individuals who are motivated to succeed are perfectionists and expect to exceed the standards of excellence, showing better organisational performance and progressing in their career in order to obtain higher salary levels (Rego, Rego & Carvalho referred to by Rego, Tavares, Cunha & Cardoso, 2005). These individuals also show a high spirit of initiative and persistence in performing tasks, showing better scores in the performance appraisal (Rego, Rego & Carvalho referred to by Rego, Tavares, Cunha & Cardoso, 2005). The importance of receiving *feedback* from others on their organisational success is also notorious (Rego, Rego & Carvalho referred to by Rego, Tavares, Cunha & Cardoso, 2005). The most successful-motivated individuals also assume money as the main indicator of success, which is reflected in better grades in courses, better jobs and, consequently, better future salaries (Rego, Rego & Carvalho referred to by Rego, Tavares, Cunha & Cardoso, 2005).

The affiliation motive, on the other hand, is defined by McClelland (1987, p.347), as "the concern with establishing, maintaining, or re-establishing a positive affective relationship with another person or persons".

According to the same author individuals, by preventing interaction and socialization with each other, individuals create the need for affiliation. This need is satisfied through the creation of affective ties that allow increasing levels of motivation at work (McClelland, 1987). However, more affiliative people who valued "warm" and affable relationships, often acting in a caring and cooperative way, when faced with threatening situations tend to act defensively and aggressively (Rego & Leite, 2003).

In relation to the attainment of management positions, individuals who are more motivated towards affiliation reveal a greater desire and effort to gain friendships and re-establish relationships, which becomes an impediment to attaining management positions and prestige, earning lower wage levels (Rego, Tavares, Cunha, & Cardoso, 2005).

With regard to cooperation and conformism in the success motive, the study of Rego and Jesuíno (2002) states that the fact that affiliative individuals prefer to maintain or re-establish affective bonds within the organisation makes them tend to be more cooperative, more conformed to the wishes of others and with better performance levels than the others. Since affiliative individuals are more cooperative, individuals, particularly females, show a greater tendency to value cooperation and, on the contrary, devalue the rationale of problem solving (McClelland, 1975 referred to by McClelland, 1987).

In   relation to the performance of joint tasks, when comparing more affinative females with less affinative males, the former are more open to asking their partners to slow down the pace of work than the latter (Walker & Heynes referred to by McClelland, 1987).

From the measurement of affiliation needs, McClelland (1987) concludes that individuals with affiliative behaviors are more sensitive to equally affiliative stimuli and are more connected to social networks compared to the rest of the group, since these individuals reveal greater social skills, with greater relational facilities, promoting interpersonal dialogue and the maintenance of relationships with other people.

When faced with a conflict situation with a stranger, individuals directed towards the affiliation motive tend to agree or disagree according to the interest aroused by the stranger. Thus, if the stranger arouses some interest in the person, he/she agrees with his/her opinion, on the other hand, if the stranger does not arouse any interest, the affiliative person tends to disagree with his/her opinion (Burdick & Burnes referred to by McClelland, 1987). In an attempt to resolve conflicts jointly, the more affluent individuals are less critical of the other group so as not to contradict them, showing a greater tendency to the avoidance strategy (Exline referred to by McClelland, 1987).

With regard to the relationship between management and affiliation, Noujaim (referred to by McClelland, 1987) states that, especially males want to maintain good relationships with their subordinates, and it is often complicated for the manager to reconcile these relationships with the complex decisions of the organisation.

Regarding the power motive, McClellland (1987, p.272) states that people motivated for power "are concerned with establishing, maintaining, or restoring power, that is, creating impact, control, or influence over the other person, group, or world at large." According to Dessler (2004) power motivation is present when individuals dominate or control the means to influence others, for example through argumentation, demand or force, conveying a direction, persuasion or punishment. According to the same author, power motivation is also visible when we are faced with a situation which is culturally accepted as an interpersonal relationship involving the superior's control over the means to dominate his/her subordinates.

People directed toward the power motive like to influence others "by making suggestions, giving their opinions and evaluations, and trying to talk about other things" (Dessler, 2004, p.292). In addition to their influence on others, these individuals like to get other people's attention by choosing activities that involve persuasion, such as teaching and public speaking, holding leadership positions, and being a member of the clergy (Dessler, 2004). Thus, these individuals show great concern with prestige, reputation and leadership positions (Rego & Leite, 2003), also denoting a remarkable verbal fluency used in persuasion with others (Pérez-Ramos, 1990).

Several studies have tried to understand the influence of managerial positions on the power motive. The study of Rego (2000), for example, revealed that managers use the power motive more often than the affiliation motive, since they need to exert power/influence over others and have to make impartial and rigid decisions which become easier when there is no creation of affective bonds. On the other hand, Steers and Porter (mentioned by Pérez-Ramos, 1990) state that employees who are motivated for power, in addition to having notorious attributes for direction (management), show high levels of performance, depending, of course, on the combination between the other needs for success and affiliation.

In short, the power motive usually takes high risks, impacts the behaviour or emotions of other individuals, seeking competitive and assertive activities in which individuals are concerned with satisfying only their needs (Rego & Leite, 2003). However, these competitive and predominantly aggressive actions are usually controlled by society, due to their destructive potential (Rego, 1995).

Finally, we conclude that individuals, in general, possess all three types of motives; however, one motive may prevail over the others, i.e., one of the motives (success, affiliation and power) will influence some individuals more than others (McClelland, 1987). Thus, an individual is more directed towards one motive than another, depending on the characteristics and motivational forces of each one. Table 3 shows us very briefly the three motives mentioned above:

Table 3 - Summary of David McClelland's motives

Reasons	The individual
Success	Seeks success through personal excellence.
	Desire to achieve high but realistic goals.
	It responds positively to competition.
	He's got initiative.
	Takes charge of tasks for which he/she can be personally responsible.
	Moderate risk-taking.
	Demonstrates a fondness for engaging with experts.
Membership	Seeks strong interpersonal relationships.
	Gather efforts to build friendships and restore relationships.
	People assume greater importance than tasks.
	Seeks acceptance from others for their opinions and activities.
Power	Seeks to influence or control others and to dominate the means to exert that influence.
	Try to take leadership positions spontaneously.
	Likes to cause shock/impact.
	Attaches importance to prestige
	You take high risks.

Source: Cunha et al (2007)

In order to finalize the topic concerning the theories of content motivation we present next a table that very succinctly presents the four theories.

Table 4 - Summary of the content theories

Maslow (1954)	Individuals are motivated through the five needs hierarchy therefore, higher level needs are only met when those of the next lower level are already satiated.
Alderfer (1972)	It is based on three needs: existence, relationship and growth. If the individual is unable to satisfy a particular need has to reinforce their efforts in order to satisfy the lower needs, leading to frustration.
Herzberg (1996)	The main factors associated with motivation are hygienic and motivational. pains, with the former preventing satisfaction and the latter leading to satisfaction.
McClelland (1961)	He argues that motivation arises when all three types of needs: achievement, affiliation and power. All individuals have these three types of needs although with prevalence of one in relation to the others.

Source: Galhanas (2009, p.8).

Chapter III: Conflict Management Strategies and Motivation

In the previous chapters it was our purpose to define and explain the concepts of conflict and motivation, and in the case of the latter concept we are particularly interested in McClelland's content theory. In the present chapter, we attempt to expose the relationship that is established in the literature (Thomas, 1992, among others) between both concepts.

The phenomenon of conflict between two or more organizations encompasses two important elements (Thomas, 1992): (1) the conflict seen as a process or a sequence of episodes that involves the individual's lived past and the behaviours perceptible from the outside; (2) the conflict determined by structural conditions that relate to the particularities of the conflicting parties and the circumstances of the surroundings. According to the same author, when we refer to the particularities of the conflicting parties, we are talking about the "*locus of* control", motivational profile, affective skills, among others. On the other hand, when we talk about group unity, compliance with deadlines and time pressures, we are referring to the circumstances of the surroundings (Thomas, 1992). Thus, it is important to study how structural conditions (motivational profile) are linked to the conflict process.

Tamayo and Paschoal (2003, p.42) state that the motivational structure of individuals is something susceptible to change, being considered a dynamic process and little stable. These same authors say that the motivational tendencies of individuals "are not always harmonious; some may be conflicting, giving rise to internal conflicts". However, and according to Deci and Ryan (1985) this conflict can be controlled. According to the same authors, extrinsic motivation occurs when certain norms of behavioural conduct are used due to the imposition of socialization agents and not due to inherent intrinsic motivation. This internalisation of norms made by external agents undergoes a process of external regulation which involves the control of the conflict and of the person who is subject to control, as well as the controller (Deci & Ryan, 1985).

Lobos (1978) states that individuals' dissatisfaction with the functions performed in the organisation is a source of conflict. According to Stratuss (referred to by Lobos, 1978), the desire for *status* when obstructed leads to conflict with other organisations. In this sense, and in this particular case, the main cause of dissatisfaction is due to the freezing of career progression and the devaluation of professionals. Argyris and Dalton (referred to by Lobos, 1978, p.348) explain that the dissatisfaction of the functions performed leads to conflict "when a unit with the same or even less *status than* another determines the performance standards for the latter".

According to Wagner and Hollenbeck (2002) one of the positive consequences of conflict is motivation. According to the same authors and, as seen earlier, the conflict can lead to the emergence of new ideas, innovation and the introduction of organizational changes that help the individual to maintain an optimal level of motivation. For their part, Kaushal and Kwantes (2006) state that the motivation for certain behaviour, when poorly clarified, leads to conflict, since the agitation and *stress of* everyday life limits individuals in terms of time, thus hindering communication between individuals who interact and preventing them from revealing what motivations led them to act in a certain way, leading them to the conflict process.

Based on some previous studies, we then associate the motivational profile with conflict management strategies, more specifically, we attempt to link the motives of power, success and affiliation with the strategies of accommodation, avoidance, imposition, commitment and integration.

Already in 1995, Urdan and Maehr (referred to by Bipp & Dam, 2014) emphasised the relationship between the success motive and conflict, as well as the importance of goal setting for norm compliance and/or the need for acceptance by others. More specifically, and according to Rego (1995), individuals more oriented towards the success motive, with the exception of managers, choose the compromise strategy as the ideal strategy for resolving their conflicts. In addition to this strategy, the most success-oriented individuals prefer to use the collaborative (or integrative) strategy which shows high concern for themselves and for others, rejecting the avoidance strategy which shows low concern for themselves and for others.

With regard to the affiliative motive, Exline (referred to by Wegner, Bohnacker, Mempel, Teubel, & Schüler, 2014) revealed that individuals who are more concerned with affective relationships have a greater tendency to avoid conflict in their work group. For example, more affinative individuals when allocated to an unfamiliar group involving group tasks avoid making decisions, thus avoiding potential arguments among the group (Exline referred to by Wegner, Bohnacker, Mempel, Teubel, & Schüler, 2014). In contrast, the study of Rego (1995) revealed that more affiliative individuals choose more cooperative strategies, preferring the use of compromise, collaboration (or integration) and accommodation strategies over the use of avoidance and imposition strategies.

In what concerns the power motive, and according to Winter (referred by Freeman, 1994) athletes when they compete with other individuals or groups tend to be more competitive than when they compete individually. On the other hand, Rego (1995) refers that individuals, when motivated for power, choose more assertive strategies, as is the case of imposition strategies or competition and collaboration, in which there is a need to satisfy their own interests, denoting a low tendency for the avoidance strategy (Rego, 1995).

Rego and Jesuíno (2002) conducted a study whose objective was to relate the strategies of accommodation, imposition, commitment, avoidance and integration with the motives of success, power and affiliation.

According to Rego and Jesuíno (2002), the success motive correlates positively with cooperative strategies, such as accommodation, commitment and collaboration, and negatively with the avoidance strategy. As regards the affiliative motive, we found that it correlates positively with the accommodation, collaboration and commitment strategies and negatively with the avoidance and imposition strategies (Rego & Jesuíno, 2002). Finally, the power motive correlates positively with the cooperative strategies and the imposition strategy and negatively with the avoidance strategy (Rego & Jesuíno, 2002).

Rego (1995) conducted a study whose objective was to cross-reference conflict management strategies with the motives of success, power and affiliation. This crossing resulted in the configurations referred in figures 3, 4, 5 and 6.

Figure 3 - Preferred zone for "Enclave Leaders

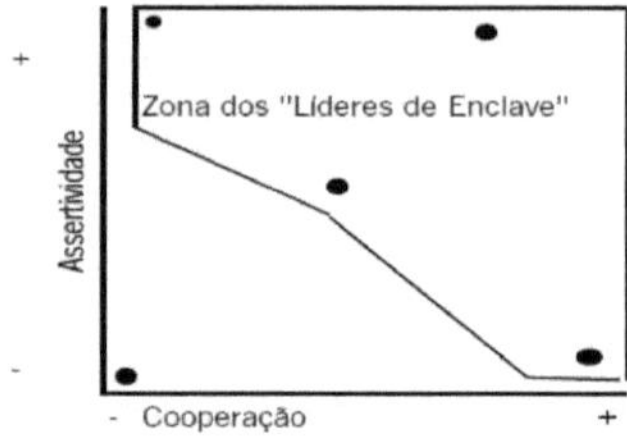

Source: Rego (1995)

Figure 4 - Preferred zone for "Successful people

Source: Rego (1995)

Figure 5 - Preferred zone for "Affiliate Normals".

Source: Rego (1995)

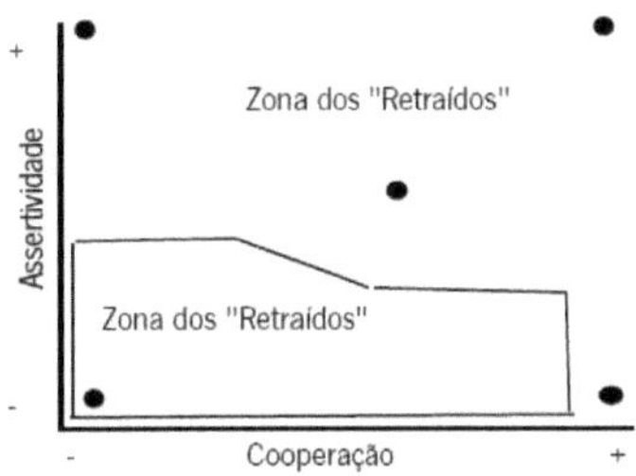

Source: Rego (1995)

According to this study, individuals belonging to the group of "enclave leaders" (figure 3) are more cooperative and assertive. This group seems to be more motivated by the motives of power and affiliation, and also shows a greater propensity for the motive of success and competition. The study by Rego (1995) suggests that "enclave leaders" preferably use the strategy of imposition, followed by the strategies of compromise and collaboration. From this group, we may also note that there is a reasonable preference for the accommodation strategy and little or no use of the avoidance strategy. Rego (1995, p.95) explains the reasonable preference for accommodation by the fact that there is "a need to obtain results, even in favour of the counterparty; and/or, then, a realistic (and calculating, in terms of future negotiations) attitude that leads them to accommodate".

In turn, individuals belonging to the "successful people" group (figure 4), as the name suggests, are individuals who show high academic performance, a high propensity for success and power and a low tendency for affiliation (Rego, 1995). In addition, these individuals, except for the accommodation strategy, use moderately the imposition, avoidance, integration and commitment strategies.

According to Rego (1995), the group designated as "normal affiliative" (figure 5) reflects the Portuguese reality, in which individuals show a greater tendency towards the affiliation motive than the remaining motives. In addition, they also show a high tendency to use the accommodation strategy.

Finally, the "withdrawn" group (Figure 6) are the least motivated individuals for both the success and power motive, denoting a moderate propensity for affiliation (Rego,

1995). Furthermore, this group denotes poor academic performance and a high preference for the avoidance strategy, i.e., individuals prefer to refuse conflict situations rather than face them.

It is based on the previous studies that the present study was architected, aiming to relate the strategies with the motives of success, affiliation and power.

Chapter IV: Characterization of the Hospitality Context

According to Castelli (1992), the hotel industry cannot be considered a mere secondary activity, since it is the fundamental element for the tourism development strategy and policy of a region or country.

Regarding the organizational structure of the hotel industry, Taraboulsi (2003) states that this sector has a diversified structure, in which each hotel unit provides its customers the most diversified services. Among them: reception (*checkin and checkout*), *conciergerie* (social reception), food and beverage, laundry and reservations.

According to Taraboulsi (2003), these services are distributed among the following departments: coordination and operations; reception and lodging; food and beverage; events; governance; and leisure and customers' well-being. According to the same author, the coordination and operations department is the most important department insofar as it is responsible for the implementation, organization, coordination and control of all hotel services.

Taraboulsi (2003) defines hospitality as the art of providing services full of prestige, friendliness, devotion and respect, determinants for customer satisfaction and enthusiasm and, especially for the humanization of service and work environment. On the other hand, being the hospitality industry a very competitive sector it is important that each participant of this industry is concerned with maintaining and achieving a prestigious position, based on quality standards (Nath & Raheja, 2001). Dickson, Ford and Upchurch (2005) share the same opinion, stating that in a competitive market such as the hospitality industry, the only differentiating aspect between companies is service excellence.

One of the main concerns of the hospitality industry is the issue of human resources. This fact is not a surprise, since the people involved in the hospitality industry are in constant conviviality with the staff and its customers. When it comes to human resources the main concerns to be taken into account are: (1) customer understanding; (2) effective capital management; (3) alignment of *stakeholder* interests; (4) use of information technology and (5) brand enhancement (Enz, 2001).

According to Spniella and Canacos (2000), people are important for any type of organization, whether it is a hotel or not, since they are an important component for the success of services, causing the organization to acquire customers and a good image through "word-of-mouth" advertising.

Regarding the hotel sector in particular, Taraboulsi (2003) refers that the triumph of this sector lies in the humanization of the environment. Through this, one obtains efficient services, well cared for spaces and, most important of all, good-natured people who, when interacting with customers, reveal a sincere smile, brand image and satisfaction for the work done (Taraboulsi, 2003). Similarly, Castelli (1992) highlights the importance of people in the hospitality industry, stating that the human element is responsible for customer hospitality, company profitability, and, depending on how customers are received, the image created about a particular region or city. By valuing people as human beings and not as objects it is possible to achieve the objectives of the organisation (Taraboulsi, 2003).

Hospitality professionals are chosen according to the characteristics and needs of organizations. Rodger (cited by Nickson, 2007, p.92) defines seven important characteristics: physical characteristics (appearance, way of speaking, and ability to carry heavy loads); skills (academic and professional qualifications and work experience); intelligence (ability to define and solve problems); special skills (talent, qualities or skills important to the position); interests (*hobbies* or leisure activities that may influence the work); disposition (sense of humour or friendliness), and finally, circumstances (family burdens and responsibilities or willingness to work more hours per day). For his part, Fraser (referred to by Nickson, 2007) states that only five requirements are needed to gain a place in hospitality. These are: creating an impact on people; having professional qualifications and experience; having innate skills and abilities; having motivation (intrinsic desire to succeed at work); and finally, having developed social skills, i.e. being competent in dealing with people.

According to Castelli (1992), and unlike other types of industries, where monotony is more present (e.g.: mass production), in the hospitality industry, the most important characteristics of professionals are dynamism and creativity, since they have greater freedom of action, being able to act in a wide variety of situations.

The regions of Algarve, Lisbon and Madeira represent about 85% of the total overnight stays of foreigners (Turismo de Portugal, 2007). Therefore, Madeira continues to be one of the tourist hotspots of choice for tourists. Therefore, hospitality is an important sector for the Autonomous Region of Madeira, as it generates employment and is responsible for a large part of external revenues. The importance of the hospitality industry for Madeira's economy, can be confirmed through graph 1, referring to the year 2012, in which the gross value added (GVA), i.e. the final result of the productive activity of a given period (Value of sales to subtract by the value of purchases) is higher in the accommodation and catering sector (hospitality sector) (Instituto Nacional de Estatística, 2014[a]).

Graph 1 - V. A. B by sector of activity in the A.R.

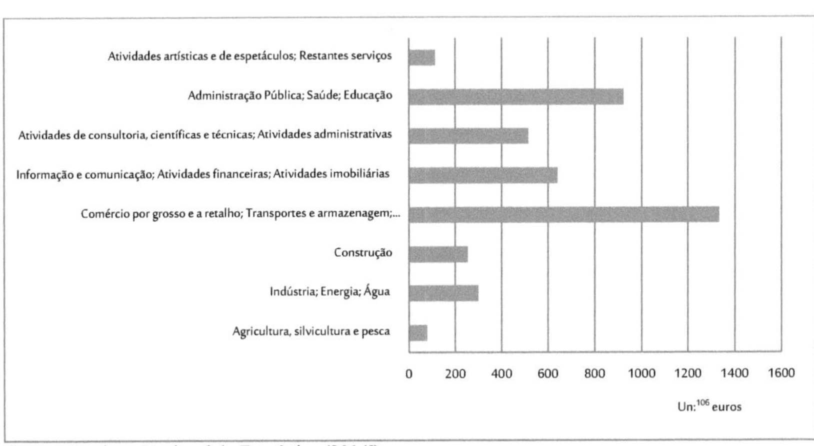

Source: Instituto Nacional de Estatística (2014[a]).

Furthermore, the importance of the hotel industry for the Region becomes clear, since and although the year 2010 was less favourable, the total revenues of this sector from 2009 to 2013 always show values equal to or greater than 250,000,000 euros (Direção Regional de Estatística, 2014).

To illustrate the above, although 2010 and 2012 were less positive years, the number of arrivals of guests in the Autonomous Region of Madeira is always equal to or greater than 840,000, and in 2013 there was an increase to 917,493 guests which is equivalent, taking into account the previous year, to an increase of about 8.23% (Regional Directorate of Statistics, 2014).

As for the number of beds, although there has been a certain oscillation in recent years, in 2013 there was a slight increase from 27,732 to 27,862 beds (Direção Regional de Estatística, 2014).

In general, there is an increase in the number of overnight stays in hotels in the region from 2009 to 2013. However, from 2009 to 2010 there is a decrease of approximately 9.16% (Direção Regional de Estatística, 2014).

The bed occupancy rate in hotels is the ratio between the accommodation capacity and the number of overnight stays. There are some oscillations in the bed occupancy rate, between the years 2009 and 2013. There is also a decrease of 4.2% in the bed occupancy rate between the years 2009 and 2010; in contrast from 2012 to 2013 there was an increase of 4.9%. It can be concluded that, since 2009, the bed occupancy rate has been increasing, which is beneficial for tourism and consequently for the Region's economy (Direção Regional de Estatística, 2014).

Regarding the number of hotel establishments in the region, in the years 2009 and 2010, the region had 201 hotel establishments, however, from 2010 this number decreased slightly, from 201 to 159 establishments in 2013 (Direção Regional de Estatística, 2014).

Hotel establishments are grouped according to the following categories: (1) pensions; (2) inns; (3) inns; (4) tourist villages; (5) tourist apartments; (6) apartment-hotels and (7) hotels. In fact and according to graph 2, most of the hotel establishments in Madeira belong to the category of hotels (Direção Regional de Estatística, 2014). It is understood that according to Castelli (1992, p.37), the concept of hotel is:

"A commercial establishment for classified accommodation, which offers furnished dwellings or apartments, for rent, either to a passing clientele or permanently, which is characterized by a rent, per week or per month, but which, with some exception, does not constitute it as a domicile. It may have a restaurant service, which operates permanently throughout the year or only in one or more seasons'.

Graph 2 - Type of hotel establishments in the Azores and Madeira

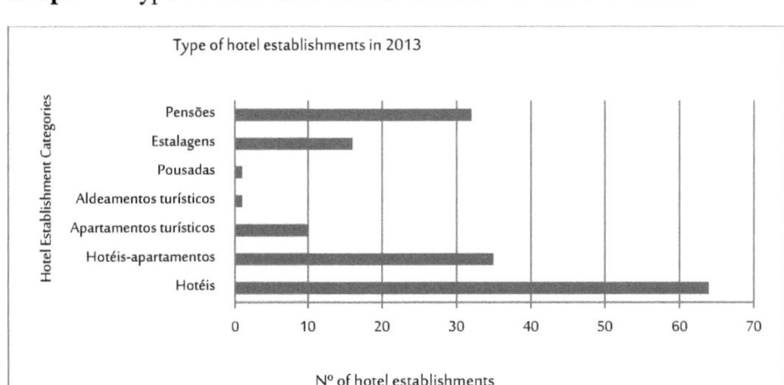

Source: Regional Directorate of Statistics (2014).

As for hotel classification as at 31 July 2013, the A.A.R. had about 68 hotels, of which 36 were 4-star hotels and 14 were 5-star and 3-star hotels. In 2013, the A.A.R. had 3,421 employees distributed among the 68 hotels, of which 1,280 belonged to 5-star hotels, 1,893 to 4-star hotels, 224 to 3-star hotels, 17 to 2-star hotels and 7 to 1-star hotels (National Statistical Institute, 2004 [b]).

In short, tourism is a fundamental sector in the economic fabric of the Autonomous Region of Madeira that has been suffering some influence of the national and international economic situation. Still, it has shown some recovery in the last two years. The activity is developed in different categories of hotel establishments, being the four star hotels in greater number, followed by five and three star hotels.

Chapter V: Methodology

According to Marconi and Lakatos (2003,p.17), scientific methodology is the scholar's introduction to "systematic and rational procedures, the basis of training" for the researcher and professional. Methodology encompasses a certain method. The term method is defined as a set of systemic and rational activities that, with greater safety and economy, enable the achievement of the objective (valid and true knowledge), detect errors and assist the scientists' decisions (Marconi & Lakatos, 2003, p.83).

This study is developed based on the quantitative methodology, since it is the most suitable for its purpose. According to Quivy and Campenhoudt (2005), the quantitative methodology is based on the frequency with which certain content characteristics appear or correlate with each other. In turn, Freixo (2011) highlights the advantages of this methodology, stating that it allows for greater accuracy and objectivity, comparison and reproduction, as well as greater generalisation and inference, which are characterised by processing and validating knowledge, generalising results, and providing the possibility of predicting and controlling events (Freixo, 2011). Having defined the methodology to be used, it is now important to define the main research objectives and the main hypotheses of the study.

5.1. Study Objectives

The objectives of this study are to identify the existing conflict strategies in the hospitality industry in the Autonomous Region of Madeira and to understand if there is any relationship between the socio-demographic variables and these conflict strategies. To this end, and in order to analyse how professionals in this industry deal with conflict, we used the questionnaire of conflict management strategies developed by Jesuíno (2003 adapted from Thomas, 1976).

It is also the aim of this study, to describe the various types of motivation of hotel companies in the Madeira Region and to verify the existence or not of a relationship between the sociodemographic variables and those same types of motivation. In order to understand which types of motivation best suit the professionals of this sector, we used the questionnaire measuring the motives of success, affiliation and power developed by Rego and Carvalho (2001, 2002). Finally, and as the main objective of this study, we intend to test the relationship between conflict management and the motivational levels of hospitality professionals.

5.2. Presentation of the Study Hypotheses

In any study, the definition of hypotheses is an important step, since it guides the research and provides a criterion for data collection that allows comparing hypotheses with reality (Quivy & Campenhoudt, 2005). For their part, Marconi and Lakatos (2003, p.126) define the hypothesis as a general statement that, by relating variables, presents a provisional solution to a given problem, has an explanatory or predictive nature, shows internal or external consistency and, finally, is susceptible to empirical verification taking into account its consequences.

In order to ascertain whether the sociodemographic variables correlate with the five conflict management strategies, we considered the following study hypotheses:

H1: Conflict management strategies will be perceived differently according to the gender of hospitality professionals.

H1a: The enforcement strategy will be perceived differently according to the gender of the professionals.

H1b: The accommodation strategy will be perceived differently depending on the gender of the professionals.

H1c: The engagement strategy will be perceived differently depending on the gender of the professionals.

H1d: The avoidance strategy will be perceived differently according to the gender of the professionals.

H1e: The integration strategy will be perceived differently depending on the gender of the professionals.

The relationship between conflict management strategies and gender was the subject of several studies. França and Lourenço (2010) found no association between intragroup conflict and gender, nor between affective conflict and gender, as did Renwick (1975) who found no differences between how men and women manage conflict. However, Dune's (1989, p.1033) study revealed that "women are more conciliatory in negotiations and are less comfortable with tasks associated with conflict management than men".

H2: Conflict management strategies will be perceived differently depending on the age of hospitality professionals.

H2a: The enforcement strategy will be perceived differently depending on the age of the professionals.

H2b: The accommodation strategy will be perceived differently depending on the age of the professionals.

H2c: The engagement strategy will be perceived differently depending on the age of the professionals.

H2d: The avoidance strategy will be perceived differently depending on the age of the professionals.

H2e: The integration strategy will be perceived differently depending on the age of the professionals.

When studying the impact of sociodemographic variables, Goel (2012) found that the variable age did not influence any of the conflict management strategies. However, Cunha (2003) in his exploratory study found that the variable age was related to the various conflict management strategies.

H3: Conflict management strategies correlate positively with the educational attainment of hospitality professionals.

H3a: The enforcement strategy correlates positively with the professionals' academic qualifications.

H3b: The accommodation strategy correlates positively with the professionals' academic qualifications.

H3c: The engagement strategy correlates positively with the educational attainment of professionals.

H3d: The avoidance strategy correlates positively with the professionals' academic qualifications.

H3e: Integration strategy correlates positively with professionals' educational attainment.

By studying these hypotheses, we intend to verify to what extent the academic level or level of education influences the conflict strategies adopted by hospitality professionals. Cunha (2003) found a positive correlation between negotiation effectiveness and academic qualifications.

H4: Conflict management strategies will be perceived differently depending on the length of time hotel employees have worked in the hospitality industry.

H4a: The enforcement strategy will be perceived differently depending on how long employees have been working.

H4b: The accommodation strategy will be perceived differently depending on how long employees have been working.

H4c: The engagement strategy will be perceived differently depending on how long employees have been working.

H4d: The avoidance strategy will be perceived differently depending on how long employees have been working.

H4e: The integration strategy will be perceived differently depending on how long employees have been working.

Marques and Cunha (2004), when analysing how sociodemographic variables correlated with the five conflict management strategies, found that individuals between 0 and 15 years of service and subjects with more than 30 years of service used the domination strategy more often when dealing with conflict.

H5: Conflict management strategies correlate positively with the professionals' employment relationship.

H5a: The enforcement strategy correlates positively with the professionals' employment relationship.

H5b: The accommodation strategy correlates positively with the professionals' employment relationship.

H5c: The engagement strategy correlates positively with the professionals' employment relationship.

H5d: The avoidance strategy correlates positively with the professionals' employment bond.

H5e: The integration strategy correlates positively with the professionals' employment bond.

According to Rahim (2001), one of the consequences of role conflict is low organizational commitment. On the other hand, and compared to the remaining strategies, the use of the integration strategy is the one most associated with high commitment to the company (Rahim, 2001).

H6: Hospitality professionals who receive training from the organization tend to learn and use a different conflict management strategy than those who do not receive training.

H6a: The enforcement strategy will be perceived differently depending on the training courses provided by the hotel.

H6b: Accommodation strategy will be perceived differently depending on the training courses provided by the hotel.

H6c: The engagement strategy will be perceived differently depending on the training courses provided by the hotel.

H6d: The avoidance strategy will be perceived differently depending on the training courses provided by the hotel.

H6e: The integration strategy will be perceived differently depending on the training courses provided by the hotel.

Since conflict is inevitable, and according to McIntyre (referred to by McIntyre 2007), for organisations to apply the most appropriate conflict management strategy, it is important to train their staff and employees. According to this same author, only through good training is it possible to effectively manage conflict, thus dealing with external competition.

H7: Conflict management strategies correlate positively with professionals with a managerial role.

H7a: The imposition strategy correlates positively with professionals with a managerial position.

H7b: The accommodation strategy correlates positively with professionals with a managerial role.

H7c: The engagement strategy correlates positively with professionals with a managerial role.

H7d: The avoidance strategy correlates positively with professionals with a managerial role.

H7e: The integration strategy correlates positively with professionals with a managerial role.

Rahim (2001) found significantdifferences between managers and the five conflict management strategies. The same author suggests that the way conflict is handled is influenced by the hierarchical relationship between the parties in conflict.

Similarly, and in order to ascertain the relationship between the demographic variables and the motivation variable, we defined the following hypotheses:

H8: The motives of success, affiliation and power will be perceived differently depending on the gender of hospitality professionals.

H8a: The reason for success tends to be different depending on the gender of the professionals.

H8b: The reason for membership tends to be different depending on the gender of the professionals.

H8c: The power motive tends to be different depending on the gender of the professionals.

War (2008), when studying the impact of the variable gender on motivation, found that this variable only reached significant values in the case of part-time workers and workers aged between 25 and 44 years. On the other hand, Rego and Jesuíno (2002), when studying motivation, found that males tended more towards the power motive and less towards affiliation, unlike females.

H9: The motives of success, affiliation and power will be perceived differently depending on the age of hospitality professionals.

H9a: The motive for success will be perceived differently depending on the age of the professionals.

H9b: The affiliation motive will be perceived differently depending on the age of the professionals.

H9c: The power motive will be perceived differently depending on the professionals' age.

Urošević, and Milijić (2012) found no significant values between the variable age and the overall satisfaction and motivation of employees. However, Luthans and Thomas (referred to by Cunha et al, 2007) argue that older workers tend to be more motivated, since over time they acquire higher hierarchical positions, with greater responsibilities and higher income.

H10: The motives of success, affiliation and power will be perceived differently depending on the educational attainment of hospitality professionals.

H10a: The motive for success will be perceived differently depending on the education level of hospitality professionals.

H10b: The affiliation motive will be perceived differently depending on the education level of hospitality professionals.

H10c: The power motive will be perceived differently depending on the education level of hospitality professionals.

Fabra and Camison (2009) studied the relationship between academic qualifications and motivation. These authors concluded that individuals with a higher level of formal education prove to be more satisfied with their jobs.

H11: The motives for success, affiliation and power will tend to be different depending on the length of service of hospitality professionals.

H11a: The reason for success will tend to be different depending on the length of service of hospitality professionals.

H11b: The reason for membership will tend to be different depending on the length of service of hospitality professionals.

H11c: The power motive will tend to be different depending on the length of service of hospitality professionals.

Ronen (referred to by Romão, 2011) studied the relationship between seniority and job satisfaction, concluding that motivation decreases in the first year of employment, stagnating over several years until it increases again.

H12: The motives for success, affiliation and power will tend to be different depending on the employment status of hospitality professionals.

H12a: The success motive will tend to be different depending on the organizational attachment of hospitality professionals.

H12b: The reason for membership will tend to be different depending on the organizational attachment of hospitality professionals.

H12c: The power motive will tend to be different depending on the organizational attachment of hospitality professionals.

Since motivation is an important factor in the workplace, it becomes essential to study its relationship with organizational attachment or commitment. According to studies by Tezergil, Köse and Karabay (2014), employees with high trust, involvement and work motivation reveal greater commitment or engagement with the organization.

H13: The motives of success, affiliation, and power will tend to differ depending on the training opportunities of hospitality professionals, with hospitality professionals who receive training from the organization tending to be more motivated for certain motives than professionals who do not receive training.

H13a: The reason for success will tend to be different depending on the training opportunities for hospitality professionals.

H13b: The reason for membership will tend to be different depending on the training opportunities for hospitality professionals.

H13c: The power motive will tend to be different depending on the training opportunities for hospitality professionals.

Several authors have tested the relationship between training and motivation, including Lacerda and Abbad (2003), Jesus (2000) and Gottfried, Fleming and Gottfried (2001).

H14: Professionals with managerial positions tend to perceive the motives of success, affiliation and power differently, and hospitality professionals with managerial positions tend to be more motivated towards success and power.

H14a: Professionals with a managerial position tend to perceive the reason for success differently.

H14b: Professionals with a managerial position tend to perceive the affiliation motive differently.

H14c: Professionals with a managerial position tend to perceive the power motive differently.

According to Rego (2000), managers are more oriented towards the power motive than non-managers. In turn, Rego and Jesuíno (2002), in a study conducted with managers, revealed in a sub-sample of these managers that this group tended to be more directed towards the power and success motive.

As seen previously, the central objective of this paper is to test the relationship between conflict management and motivation. Thus, we start from the following null hypothesis:

H0: Hospitality professionals do not tend to correlate conflict management strategies with motivation.

H15: Hospitality professionals tend to positively correlate the strategies of imposition, accommodation, compromise, avoidance, and integration with the motives of success, affiliation, and power. Several authors, in previous studies have tested the relationship between various conflict management strategies and motivational patterns, among them: McClelland (1987), McAdams (1992) and Murray (1938).

5.2. Data Collection Instrument

The instrument used for data collection was the questionnaire. The questionnaire includes questions that address the problems to be investigated. In this particular study, questionnaires were used because: a) the relationship between conflict management and motivation seems more valid when the instrument used is the questionnaire; b) the questionnaire has reasonably high levels of reliability when compared to the projective test; c) the questionnaire is an instrument with greater practicability than the projective test (Rego, 2000). According to Barañano (2004), questionnaires are a quantitative method that belongs to the positivist paradigm. These are done through information sharing in which, from a representative sample, important data for research are collected (Barañano, 2004). In turn, Fortin (2000) states that the questionnaire, as its name indicates, aims to question a set of representative individuals of a given population, in order to deduce and generalize, as from a particular case.

In this research, three different questionnaires are used depending on the intended purposes: the sociodemographic and professional questionnaire (Annex II); the questionnaire concerning conflict management strategies (Annex III); and finally, the questionnaire measuring the motives for success, affiliation and power (Annex IV). Normally, the questionnaires applied, may be of two types: open or closed response. Open-response questionnaires give the respondent the opportunity to express his/her thoughts and opinion, whereas in closed-response questionnaires, the subject is limited to answering the range of questions provided, without being able to modify or argue them. The questionnaires used in this study are closed-ended questions, since they are more appropriate to the study's objective, are the most commonly used, allow for greater uniformity of answers and greater ease of processing (Gil, 2008).

The sociodemographic questionnaire was used to characterise the demographic aspects of the sample, gender and age, as well as aspects related to the professional situation. This questionnaire consists of 8 clear and direct questions.

In   order to analyse how individuals face conflicts in the hospitality sector, the questionnaire on "conflict management strategies" referenced by Jesuíno (2003 adapted from Thomas, 1976) was used. This questionnaire contains 20 questions that are distributed by the five conflict management strategies: imposition (items 1,6,11 and 16); commitment (items 3,8,13 and 18); integration (items 5,10,15 and 20); accommodation (items 2,7,12 and 17) and avoidance (items 4,9,14 and 19).

The scale used was the Likert scale which measures attitudes and, according to Gil (2008), is used for respondents to express their agreement or disagreement in relation to each item. Thus, and in order to assess each response, a higher value is assigned to the most favourable response and vice versa. Therefore, with regard to the classification, it should be clarified that an employee with a high score in the imposition strategy is more interested in satisfying his/her own interests, neglecting the interests of others. An employee with a high score in the commitment strategy, on the other hand, expresses the satisfaction of his/her needs through a set of sacrifices and compromises made by stakeholders. On the other hand, a hospitality professional with a high score in the integrative strategy holds a greater concern for his or her interests and those of others. In case the employee gets a higher rating in the accommodation strategy, it means that there is a greater concern for satisfying the interests of others. Finally, an employee with higher rating in the avoidance strategy ignores or neglects the interests of both parties.

The answers to the 20 questions are presented on a Likert scale, with item 1 being "rarely", item 3 "sometimes", and item 5 "always" (Table 5). The scale is rated based on the sum of the questions of each of the 5 conflict management strategies, and may vary between 1 and 20. The mean value of each strategy is divided by four to obtain the average value.

Table 5 - Questions related to conflict management strategies

1. I defend my position with tenacity.	**Imposition**
6. I try to identify what is wrong with the other person's position.	
11. I like to win an argument.	
16. I find it difficult to admit that I am wrong.	
3. I try to reach a compromise acceptable to both parties.	**Commitment**
8. I negotiate to get some of what I propose.	
13. I look for a middle ground to resolve disagreements.	
18. I emphasize the advantages of "give and take".	
5.I try to examine the problems together in a comprehensive manner.	**Integration**
10. I openly share with others the information available to me in order to resolve the points of contention.	
15. I encourage an open exchange about disagreements and problems.	
20. I present my position as being only one point of view.	
2. I try to put the needs of others above my own.	**Accommodation**
7. I seek to promote harmony.	
12. I go along with the suggestions of others.	
17. I try to help others to avoid "loss of posture".	
4. I try not to get involved in conflicts.	**Avoidance**
9. I avoid opening discussions about controversial aspects.	
14. I keep what I feel to myself in order to avoid misunderstandings.	
19. I encourage others to take initiative in resolving the controversy.	

The second questionnaire used was the "questionnaire measuring the motives of success, affiliation and power" by Rego and Carvalho (2001, 2002) which aims to assess the types of motives that are most associated with each individual. This questionnaire by Rego and Carvalho (2001, 2002) resulted from a study carried out with university students, with the purpose of understanding the motivational profiles of the subjects in question. This measurement instrument consists of 18 questions corresponding to the different motives. Items 1, 4, 7, 10 and 13 correspond to the success motive, items 2, 5, 8, 11, 14, 16 and 18 to the affiliative motive, and items 3, 6, 9, 12, 15 and 17 to the power motive. Hospitality professionals who scored highly on the success motive have a high spirit of initiative, are by nature more competitive, and prefer more difficult tasks. Professionals scoring higher in the affiliative motive value affective relationships, are more cooperative, and have a high level of performance. Finally, employees with high scores in the power motive like to dominate and influence others, prefer to assume leadership positions and high *status*. Thus, the motives are distributed into three dimensions, and to each of them corresponds a different set of questions. The answers to the 18 questions are presented according to a Likert scale from 1 to 7, where item 1 is "never", 4 "sometimes" and 7 "always" (Table 6). The scale is rated based on the sum of the questions for each of the 3 reasons. To obtain the mean value of each reason, the sum value obtained is divided by the number of items of each reason.

Table 6- Questionnaire of the three reasons.

1. I like to constantly improve my personal skills.	
4. I strive to improve on my previous results.	
7. I like to know whether or not my work has been well done, so to do better in the future.	**Success**
10. At work, I try to do better and better.	
13. I try to do my work in an innovative way.	
2. I like to be supportive of other people, even if they are not of my relationships.	
5. I feel satisfaction when I see that a person who has asked me for help be happy with my support.	
8. If you had to dismiss a person, you would seek above all to understand her feelings and support her in any way I could.	**Membership**
11. At work, I like to be a kind person.	
14. I feel satisfied working with people who like me.	
16. At work, I pay close attention to the feelings of others.	
18. I am concerned when I feel that I am contributing in any way to the malaise of relationships at work.	
3. I have a secret desire to get people's attention.	
6. I insist on a certain opinion just to "not give the arm to be twisted".	

9. I have arguments with others because I tend to insist on what I think it should be done.	
12. I try to relate to influential people.	**Power**
15. If I can call on people for my team's work, I look for those that allow me to exert more influence.	
17. When I participate in a social gathering, I use the opportunity to influence others and get their support for what I want to do.	

5.3. Data Collection Procedure

Data were collected during the first semester of the 2014/2015 school year. First, an informal contact was made with the people in charge of the various hotel establishments to find out about the possibility and availability for the completion of the questionnaires. In this first approach, the questionnaires were made known, explaining the procedures and objectives of the respective research, as well as the method to be used in data collection. After contacting the various managers of the hotel establishments, we agreed on a day for the delivery and collection of the questionnaires in advance, so as to ensure the availability and cooperation of hotel professionals. In order to formalize the request for authorization to deliver the questionnaires, a letter requesting authorization (Appendix I) to carry out this type of study, explaining the importance and main objectives of the study, was delivered to the managers of the various hotel establishments. A copy of the questionnaire was attached to this formal authorization letter.

Subsequently, and with the proper authorization, the questionnaires were distributed in paper format and personally delivered on the agreed-upon day. When the questionnaires were delivered, all the necessary information was provided, i.e., information was provided about the personal identity, advantages and objectives of the study, as well as the guarantee of data confidentiality and participant volunteering.

The questionnaire was applied in 2015, between January 10 and 31, since, for the present study, the cooperation of approximately 27 hotels (2, 3, 4 and 5 stars) was necessary, being distributed on different days. Additionally, the instructions for filling out the questionnaires were provided in the questionnaire header.

The questionnaires as previously mentioned, were delivered personally and in paper format. In all, 410 questionnaires were distributed, however, only 349 of the total checked were received.

For this study, and in order to facilitate data processing, we used the SPSS *(Statistical Package for the Social Sciences)*, Windows version 22. According to Maroco (2011), this *software program* is possibly the most widely used in the social sciences and humanities, biomedicine, business sciences and engineering.

5.4. Population and Sample

According to Bussab and Morettin (2002), the target population is a statistical term used to designate a set of components, or results of the research. According to these same authors, this concept is distinct from the sample, since the latter refers only to a part of the population.

The target population of the study consisted of hotel professionals from some establishments in the Autonomous Region of Madeira (27 hotels). The sample participants were randomly selected, depending only on their availability and presence at the time and day indicated. Thus, the sample is of convenience, to the extent that data collection depends only on the accessibility and presence of the hotel industry employees (Gil, 2008).

Chapter VI: Presentation and Analysis of Results

In this topic, we initially present the sample and then the analysis of the results concerning the descriptive analysis and inferential analysis. Since we describe a reality and test this same reality through the formulation of explanatory hypotheses, we used the descriptive and correlational type of study. Thus, descriptive research "aims to describe the phenomenon or the characteristics of a given population" (Alves, Dacoreggio, Becker & Teixeira, 2008,p.3). Marconi and Lakatos (2003) also add that the descriptive study aims to describe a given universe or phenomenon or correlate variables, and this type of study often uses standardized techniques for data collection. As for the correlational study, it is characterised by using non-manipulable objectives and phenomena, with the purpose of exploring the relationships between them (Lukas & Santiago; Borg and Gall referred to in Coutinho, 2008).

6.1 Characterisation of the sample

From the universe of hotels in the Autonomous Region of Madeira, we collected a sample of 349 employees who currently work in 27 hotels belonging to the categories of 2, 3, 4 and 5 stars, having participated in the study 2 hotels belonging to the 5-star category, 18 4-star hotels, 6 3-star hotels and 1 2-star hotel.

In this study, 33.2% of the participants reported being aged between 30 and 39 years. This age group is composed of 116 young adults, which is equivalent to the majority age group.

The respondents of this study are mostly female, comprising 60.2% of the sample (n=210), prevailing significantly over males, which correspond only to 39.8% of the participants (n= 139).

With regard to the variable length of professional experience, the mean is 17.61 years, of which 38.4% (n=134) had between 10 and 19 years of service. It should be noted that subjects had at least 1 year of professional activity and a maximum of 46.

In terms of academic qualifications, secondary education reached 54.20% (n=189) of the sample, followed by lower secondary education with 14.60% (n=51). A total of 13.50% (n=47) of the group under study had a Bachelor's degree.

The data reveal that most of the professionals in the sample, more precisely 74.2% are permanent employees, while 22.1% are contract employees. About 3.7% belongs to the category "other" and includes professionals with trainee status.

Of the total participants in the study, 56% chose to work in the hotel due to a liking or interest in the   area and 16% due to knowledge, or experience in the area, gained previously.

In terms of attending training courses provided by the hotel, 73.9% (n=258) of the respondents answered affirmatively. The remaining 26.1% (n=91) responded negatively.

With regard to management positions, 66.8% of the sample members have no management positions, while the remaining 33.2% of the individuals have leadership positions. The distribution of the sample is unbalanced due to the composition of the work teams. Most hotels are organized by departments led by a manager, who is responsible for coordinating the activities of his subordinates.

6.2 Descriptive Analysis

In this topic, we will first analyse the reliability and descriptive analysis of the conflict management scale, and then we will also analyse the reliability and descriptive analysis of the motives scale.

Initially, the reliability of the conflict scale was tested. Thus, according to Hayes (1998), reliability indicates the extent to which the observed results are related to the true values, i.e., to understand the extent to which it contains no random error. Reliability can thus be measured in three different ways: retest, equivalence and internal consistency (Hayes, 1998). The internal consistency of the scale of conflicts and motivation was assessed using Cronbach's alpha, which is used to test the extent to which items are correlated (Cortina, 1993). In turn, Streiner (2003) defines Cronbach's alpha as the average of the correlations of the various items belonging to a given instrument. Thus, the results of the Cronbach's alphas concerning the conflict management scale are presented in Table 7:

Table 7 - Cronbach's alphas of the conflict scale

	Cronbach's alpha before removal of items	N items	Cronbach's alpha after removal of items	N of items
Imposition	0,136	4	,552	2
Commitment	0,572	4	,646	3
Integration	0,461	4	,573	2
Accommodation	0,484	4	,578	2
Avoidance	0,525	4	,679	2

When analysing Table 7, we found that each of the five conflict strategies was composed of 4 items and had Cronbach's alphas always below 0.60 (according to Maroco & Garcia-Marques (2006) in some social sciences settings, an α is only considered acceptable when it reaches a value of 0.60). Therefore, we chose to remove some items so as to increase the alphas and, consequently, the reliability of the conflict strategies. Thus, in the imposition strategy, we removed items 1 ("I defend my position tenaciously") and 6 ("I try to identify what is wrong with the other's position"); in the commitment strategy, we only removed item 8 ("I negotiate to obtain part of what I propose"); in the integration strategy, we removed items 15 ("I encourage an open exchange about disagreements and problems") and 20 ("I present my position as being only one point of view"); In the accommodation strategy, we eliminated items 2 ("I try to put the needs of others above my own") and 12 ("I go along with the suggestions of others"); and finally, in the avoidance strategy, we removed items 14 ("I keep what I feel to myself in order to avoid misunderstandings") and 19 ("I encourage others to take the initiative in resolving the controversy"). After the removal of the items, we found that the new values of Cronbach's alphas range between a minimum of 0.552, in the imposition strategy, and a maximum of 0.679, in the avoidance strategy. Thus, the avoidance strategy (α=0.679) has the highest alpha, followed by the commitment strategy, with an alpha of 0.646, and both strategies have an acceptable internal consistency (Maroco & Garcia-Marques (2006) state that α=0.60 is considered acceptable in some social science settings). However, the strategies of imposition, integration and accommodation present low values of Cronbach's alphas, since they are lower than 0.60. However, we decided to keep these strategies, despite their low Cronbach's alphas. The reason that led us to make this choice is related to the fact that it is important for this research to continue considering the concept of conflict conceptually linked to the five strategies on which our hypotheses are anchored. In Rego and Jesuíno's study (2002), alphas lower than 0.60 were also considered, namely for the imposition strategy (α=0.47).

Next, we analysed the answers to the questionnaire on conflict management strategies (Jesuíno, 2003 adapted from Thomas, 1976).

Of the 2 items that make up the imposition strategy, and according to Chart 3, it can be seen that respondents chose the option "sometimes" to answer item 11 and "rarely" to answer item 16 (41%; n=143). In this conflict strategy, about 49% of the subjects admit that sometimes they like to win an argument (item 11). Only in item 16, 41% of the sample participants say that they rarely have difficulty admitting that they are wrong.

Graph 3- Relative frequencies of the imposition dimension

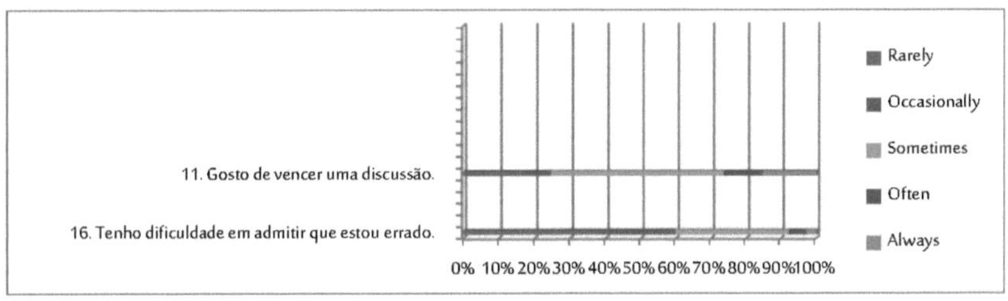

In the compromise strategy and according to Graph 4, for items 3 and 18 most respondents answered the option "always", while the option "sometimes" was more frequent in item 13. Of the sample collected, 44.7% always try to reach a compromise acceptable to both parties involved in conflict (item 3), as well as 40.4% of the subjects said they always value the advantages of giving and receiving (item 18). Sometimes, 36.1% of the subjects state that they seek a middle ground to resolve conflicts (item 13).

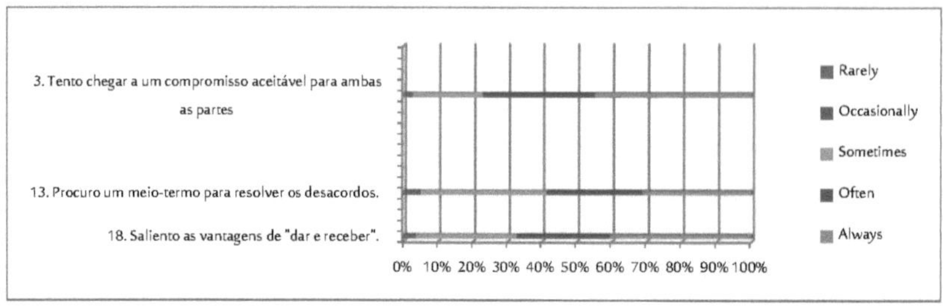

In the integration strategy (Graph 5), approximately 34.8% stated that they sometimes try to analyse problems jointly and thoroughly (item 5). In statement 10, 41.4% of respondents chose to answer that they always share with others some information they have in order to solve the disagreement.

Graph 5 - Relative frequencies of integration.

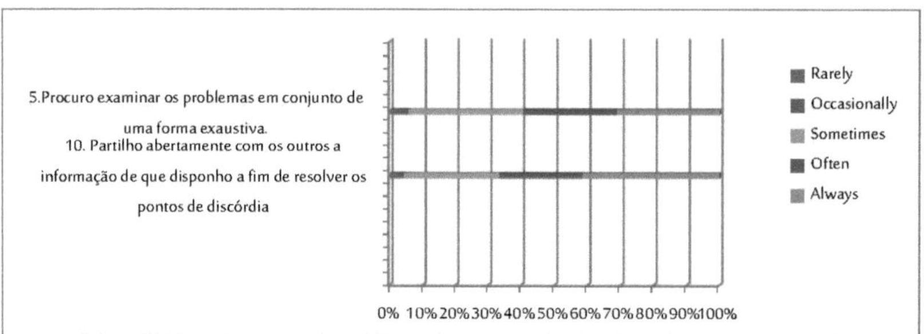

With regard to the accommodation strategy and according to Chart 6 in relation to statement 7, approximately 64.5% (n=225) of the respondents answered "always", while the option "sometimes" was more frequent in statement 17. Thus, we can also add that most respondents sometimes prioritise the interests of others, even superimposing the objectives of others over their own.

Graph 6- Relative frequencies of accommodation

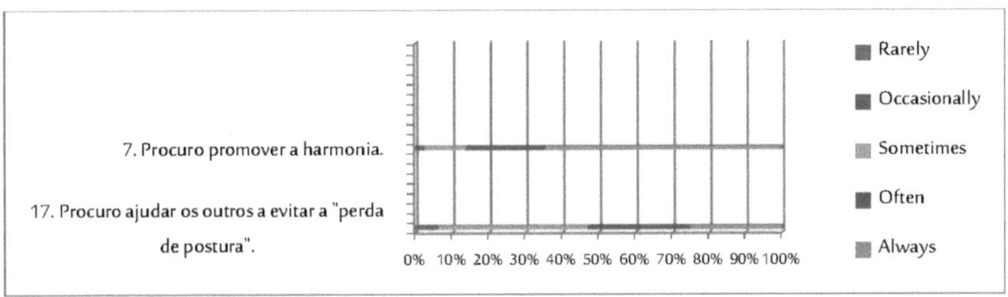

Finally, and according to Graph 7, regarding the avoidance strategy, the option "always" was more frequent in items 4 and 9. In general, in the avoidance strategy, we found that in both items (4 and 9), most hospitality employees always like to avoid discussions and conflicts.

Chart 7- Relative frequencies of the avoidance strategy.

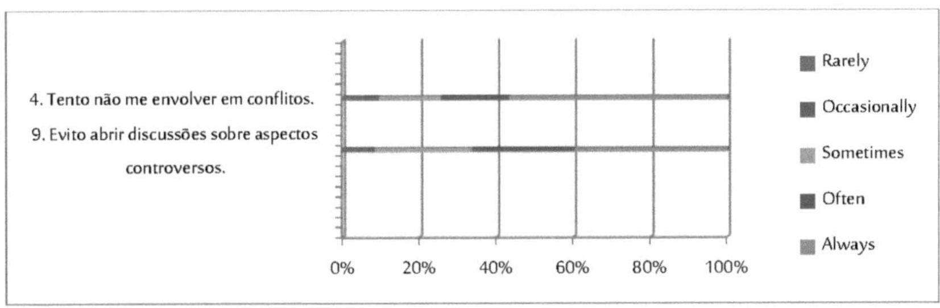

With regard to the minimum and maximum values, with the exception of the imposition and avoidance strategies, in which the participants' answers ranged between 1 (rarely) and 5 (always), the accommodation, commitment, and integration strategies ranged between 2 (occasionally) and 5 (always). With regard to the mean variable, we found that the highest value belongs to the accommodation strategy, with a mean of 4.09 (SD=0.69), followed by avoidance and commitment, with a mean of 4.04 (SD=0.96) and 4.01 (SD=0.70) respectively. At the lowest values, we found the integration strategy with a mean of 3.92 (SD=0.77) and the imposition strategy with a mean of 2.55 (SD=0.91). Based on the analysis of the standard deviation, we found that there is no significant variation or dispersion between the values, i.e. the variation in the answers does not deviate much from the mean. The results are shown in the following table:

Table 8 - Min, Max, Mean and Standard Deviation of the Conflicts scale

	Minimum	Maximum	Average	Standard Deviation
Imposition	1	5	2,55	0,91
Commitment	2	5	4,01	0,70
Integration	2	5	3,92	0,77
Accommodation	2	5	4,09	0,69
Avoidance	1	5	4,04	0,96

According to Table 8, the best mean corresponds to the accommodation strategy, with a mean of 4.09, meaning that individuals tend to be more oriented towards the accommodation strategy, where there is a low concern in satisfying their own interests and a high satisfaction in satisfying the interests of others. Also through the analysis of Table 8, it was found that the second best mean belongs to the avoidance strategy revealing a low concern of hospitality employees with themselves and with others.

According to McIntyre (2007), what usually happens is that people do not feel comfortable talking and more often use avoidance and accommodation strategies. This may result in a *deficit of* participation and active intervention by employees in their organisations with the aim of resolving conflict situations so that they do not feel harmed.

Once the internal consistency and the analysis of the scale of conflicts were tested, the reliability and descriptive analysis of the scale of motives were verified.

In the study conducted by Rego and Carvalho (2002), the Cronbach's alpha values for the Portuguese population were 0.79 for the success motive, 0.76 for the affiliation motive and 0.71 for the power motive. Thus, and according to the data collected in this study, the following results were obtained regarding Cronbach's alphas (Table 9):

Table 9- Cronbach's alpha of the motives scale

Cronbach's alphas	Number of items	
0,861	5	Success
0,838	7	Membership
0,833	6	Power

The alphas, as we can see, vary between 0.833 and 0.861. Through the table, we found that the success motive has the highest alpha value (0.861), followed by the affiliation motive with an alpha of 0.838. The power motive reveals a lower alpha, being 0.833. In comparison to the study conducted by Rego and Carvalho (2002), these Cronbach's alphas were slightly higher, which shows a good internal consistency and, consequently, a good reliability of the scale of motives.

In order to understand the motivational profiles present in the sample, we performed a descriptive analysis of the results obtained through the questionnaire measuring the "motives for success, affiliation and power" developed by Rego and Carvalho (2001, 2002). The questionnaire in question consists of 18 statements, which are equivalent to different types of motivation. Respondents were given 7 response options, from item 1 "never" to item 7 "always".

For each type of motivation, we analysed the answers obtained by the sample participants. With regard to the success motive and according to Chart 8, the most frequent answer was "always", and this was the option used to answer the 5 statements integrating this type of motive.

Graph 8 - Frequencies regarding the reason for success

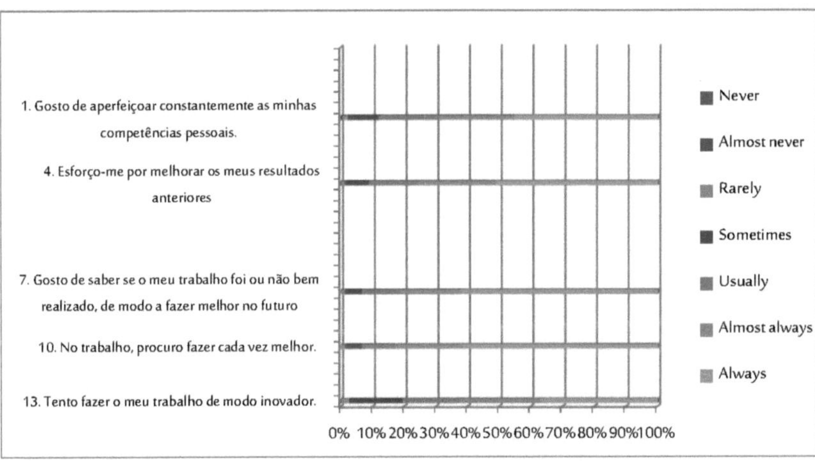

In this sense, it should be noted that in the dimension related to the reason for success, most employees say they have experienced the feelings related to the 5 statements proposed, revealing a great tendency to seek success. Approximately 45.3% of those surveyed always like to improve their personal skills (item 1). Similarly, 51% of the respondents always strive to improve their previous results (item

4). About 62.5% of the individuals always like to know whether or not their work was well done, with a view to improving it in the future (item 7). In addition, it can be stated that 61.6% of the individuals refer that they are always concerned with doing their work better and better (item 10). Finally, and with a more tenuous percentage, about 36.8% of respondents often try to do their work in an innovative way (item 13).

Of the 6 statements belonging to the power motive, the option "never" was the most frequent answer, with the exception of item 12 in which 27.2% (n=95) of the individuals answered "sometimes" (Graph 9).

Graph 9 - Relative frequencies of the power motive.

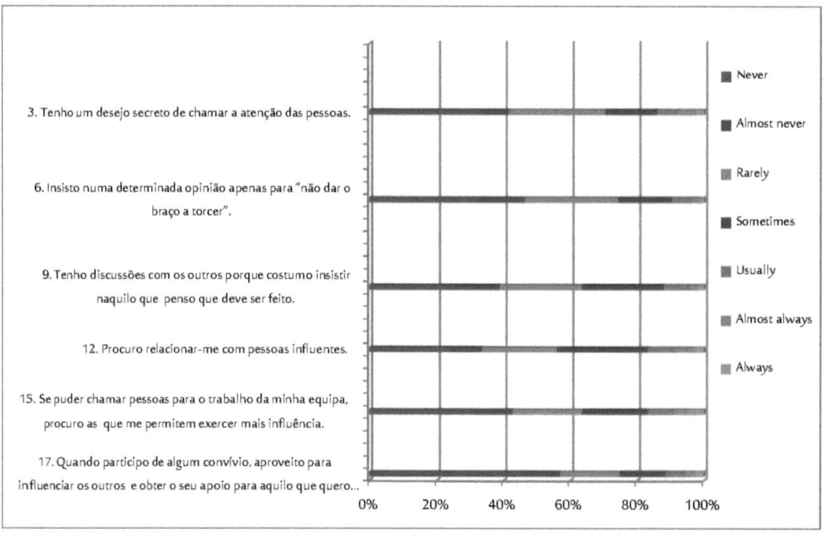

Thus, of the 6 answered items, most individuals stated that they had never experienced the situations described in this dimension, denoting a low level of power. In this way, about 33.7% of hotel employees say they never want to draw other people's attention (item 3). Of the individuals who cooperated in the study, 32.7% say they never insist on a certain idea, just to "not give the other person's attention" (item 6). About 25.2% of employees say they never argue with others for insisting on what they think (item 9). Except for the previous items, in which individuals mostly opted for answer 1 (never), about 27.2% of the individuals consider that they sometimes seek to relate to influential people (item 12). 30.9% of respondents ensure that they never call to their work team people who are influential (item 15). Finally, about 45.8% of the participants affirm that they have never taken advantage of a social gathering to influence or ask for support to get what they want (item 17).

Finally and according to graph 10, in the affiliation motive most subjects chose to answer option 5 (always) to the 7 questions pertaining to this motive.

Graph 10 - Relative frequencies of the affiliative reason.

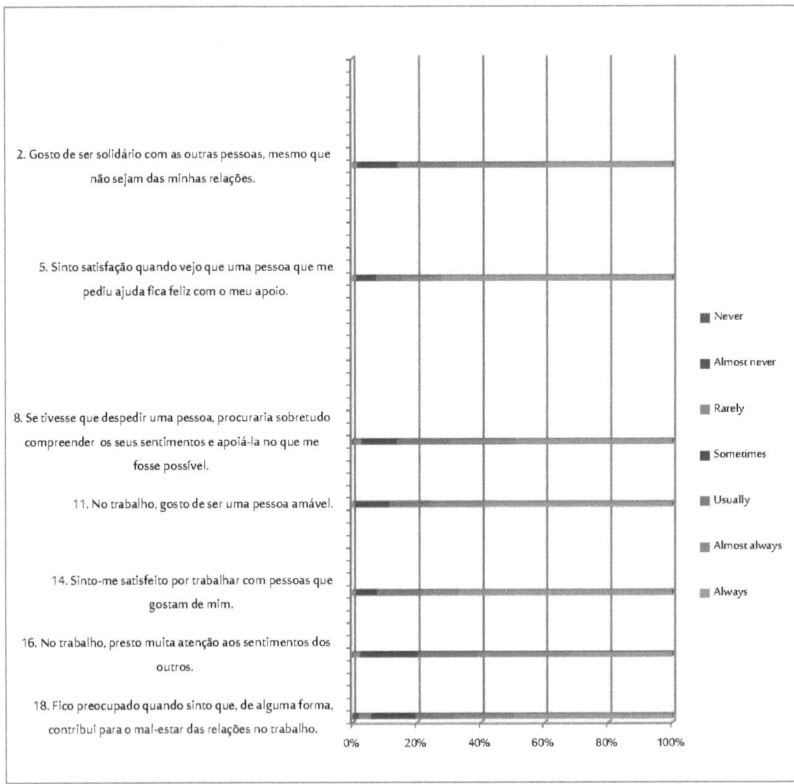

The sample reveals that most individuals identify with the 7 situations described in this reason, denouncing good levels of affiliation. We can also refer that 39.5% of the employees affirm that they are always solidary, even if that person is not of their relations (item 2). Around 72.2% of the subjects always feel great satisfaction when the person who asked for help is happy with their support (item 5). Similarly, 49.4% of the respondents recognize that in a situation of dismissal they would always try to understand the feelings and support the person who was dismissed (item 8). Of the participants in the study, 58.7% consider themselves always very kind in the workplace (item 11). 66.8% of employees report that they always feel very satisfied to work with people who like them (item 14). In the workplace, 32.1% of the subjects are always attentive to the feelings of others (item 16). Finally, 49.6% of the subjects always feel very concerned when they realize that they have contributed to the unease of work relationships (item 18).

As   regards the mean scores, the success reason varies between 2 and 7, while the affiliation reason varies between 4 and 7 and the power reason varies between 1 and 7. With regard to the mean scores per reason, we found that the highest mean score belongs to the success reason with a value of 6.13 (SD=0.86), followed by the affiliation reason with a value of 6.08 (SD=0.82) and the power reason with a mean score of 2.80 (SD=1.17). These results are shown in the following table:

Table 10- Min, Max, Mean and Standard Deviation of the motives scale

	Minimum	Maximum	Average	Standard Deviation
Success	2	7	6,13	0,86
Membership	4	7	6,08	0,82
Power	1	7	2,80	1,17

The exploratory study conducted by Rego, Tavares, Cunha, and Cardoso (2005) on higher education students reveals that they tended to be more motivated towards success (mean=6) and less motivated towards affiliation (mean=5.8) and power (mean=3.6). Also in this study, hospitality employees seem to tend to be more motivated towards success and less towards affiliation and power. However, recall that there is a great dispersion in the averages obtained in relation to the power motive.

6.3 Inferential Analysis

Loureiro and Gameiro (2011, p.155) define hypothesis testing as:

> "a set of statistical inference techniques that assess the probability that certain deviations, differences, or relationships recorded in observed distributions are due to sampling chance or actual existence in the population, based, of course, on observations made in samples."

The hypothesis tests used in this study were non-parametric tests. The reasons why we chose this type of test are related to the fact that we were unable to obtain normality and homogeneity for the sample, as we will see later on. According to Maroco (2011), non-parametric tests are traditionally used in the social sciences and humanities as an alternative to parametric tests when the normality and homogeneity of the study are not verified. Thus, in the area of psychology, non-parametric tests are often used, since, according to Maroco (2011), variables in this area are not at all normal, since human beings do not behave randomly.

Non-parametric tests allow the use of simple calculations, categorical or ordinal data, and are also less sensitive to measurement errors and more efficient than parametric tests when they are not close to the normal distribution (Filho, Viola & Borges, 2010).

To test the normality of our sample, the Kolmogorov-Smirnov test was used. This test is used for large samples and to assess whether a set of observations belongs to a normal population or not (Lilliefors, 1967). This test is also used in the case where the mean and variance are not specified (Lilliefors, 1967). Thus, normality and the hypotheses concerning conflict management strategies were initially tested.

In our case, and in relation to the hospitality professionals of the R.A.M., we found that the conflict management strategies do not approach the normal distribution, thus we rejected the null hypothesis (Ho: X has a normal distribution) since our $p<0.05$. According to Maroco (2011) a sample approaches the normal distribution when it assumes significance values greater than 0.05 ($p>0.05$). By observing Table 11, we found that the significance level is lower than 0.05, confirming the lack of homogeneity of our sample.

Table 11 - Kolmogorov-Smirnov normality test for the variable conflict management.

		Imposition	Accommodation	Commitment	Avoidance	Integration
N		349	349	349	349	349
Normal parametersa,b	Average	2,55	4,09	4,01	4,04	3,92
	Error Deviation	,913	,693	,697	,965	,769
Most Extreme Differences	Absolute	,150	,167	,104	,167	,154
	Positive	,131	,157	,094	,159	,153
	Negative	-,150	-,167	-,104	-,167	-,154
Test statistics		,150	,167	,104	,167	,154
Significance Sig. (2 extremities)		,000c	,000c	,000c	,000c	,000c

a. The distribution of the test is Normal.
b. Calculated from data.
c. Lilliefors' correction for significance.

Having tested the normality of conflict management strategies, we now check the various study hypotheses regarding the scale of conflicts.

With regard to the gender variable and to test Hypothesis 1 (H1a: the imposition strategy will be perceived differently depending on the gender of professionals; H1b: the accommodation strategy will be perceived differently depending on the gender of professionals; H1c: the commitment strategy will be perceived differently depending on the gender of the professionals; H1d: the avoidance strategy will be perceived differently depending on the gender of the professionals; H1e: the integration strategy will be perceived differently depending on the gender of the professionals) we used the Mann-Whitney test. This test is used to test whether two independent samples come from the same or similar populations (Filho, Viola, & Borges, 2010). Additionally, and according to Iochida and Castro (2001), the Mann-Whitney test is a test used for a non-normal type of quantitative variable, and is also used for the comparison of nominal variables.

Table 12 - Mann-Whitney test for gender/management conflicts

	Female			Male				
	Average	Standard deviation	Median	Average	Standard deviation	Median	Z	Significance Sig. (2 extremities)
Imposition	2,6	0,9	2,5	2,5	0,9	2,5	-0,305	0,76
Accommodation	4,1	0,7	4,0	4,1	0,7	4,0	-0,59	0,55
Commitment	4,0	0,7	4,0	4,0	0,7	4,0	-0,46	0,65
Avoidance	4,1	1,0	4,5	3,9	1,0	4,0	-1,57	0,12
Integration	3,9	0,8	4,0	3,9	0,8	4,0	-0,06	0,95

a.   Cluster Variable: Gender of the participant

Table 12 shows that the five conflict management strategies are not related to the individuals' gender, as they have p-values greater than 0.05, thus rejecting H1, i.e. the gender variable does not influence conflict management strategies.

With regard to age and to test hypothesis 2 (H2a: the imposition strategy will be perceived differently depending on the age of the professionals; H2b: the accommodation strategy will be perceived differently depending on the age of the professionals; H2c: the commitment strategy will be perceived differently depending on the age of the professionals; H2d: the avoidance strategy will be perceived differently as a function of the professionals' age; H2e: the integration strategy will be perceived differently as a function of the professionals' age), we used the Kruskal-Wallis test which is used to check whether two or more samples come from the same population or from different populations or if the sample taken starts from the same distribution (Maroco, 2011, p.317). According to Iochida and Castro (2001), the Kruskal-Wallis test is used to compare ordinal or non-normal variables for more than two groups, and is also considered a generalization of the Mann-Whitney test.

Table 13 - Kruskal-Wallis test for age/management conflicts

	Imposition			Accommodation			Commitment			Avoidance			Integration		
Age	Average	Standard deviation	Median	Average	Standard deviation	Median	Average	Standard deviation	Median	Average	Standard deviation	Median	Average	Standard deviation	Median
Up to 30 years old	2,9	1,1	3,0	3,9	0,7	4,0	3,8	0,7	3,7	3,8	0,9	4,0	3,8	0,8	3,5
30-39 years old	2,4	0,9	2,3	4,2	0,7	4,0	4,1	0,6	4,0	4,2	0,9	4,5	4,0	0,7	4,0
40-49 years old	2,6	0,8	2,5	4,0	0,6	4,0	4,0	0,7	4,0	4,1	0,9	4,0	3,9	0,7	4,0
50-59 years old	2,5	0,9	2,5	4,1	0,8	4,0	4,1	0,7	4,0	3,9	1,1	4,0	3,9	0,9	4,0
60-70 years old	2,5	0,9	2,5	3,9	0,6	4,0	4,0	0,8	4,0	4,1	0,8	4,5	3,8	0,7	4,0
Chi-square	12,16			8,51			4,41			5,59			2,83		
Significance Sig	0,02			0,07			0,35			0,23			0,59		

When comparing the means between the different age groups and the five conflict management strategies, and with the exception of the imposition strategy, we found that there are no significant variations between the means, leading our significance level to be higher than 0.05. Table 13 shows that the imposition strategy has significance values lower than 0.05, thus we accepted hypothesis H2a, which refers that there are significant differences between the variable age and the imposition strategy. However, both the accommodation and commitment strategies and the avoidance and integration strategies, when analysed with the age variable, assume significance values higher than 0.05, thus rejecting hypotheses H2b, H2c, H2d and H2e.

With regard to academic qualifications and in order to test Hypothesis 3 (H3a: the imposition strategy correlates positively with the professionals' academic qualifications; H3b: the accommodation strategy correlates positively with the professionals' academic qualifications; H3c: the commitment strategy correlates positively with professionals' educational attainment; H3d: the avoidance strategy correlates positively with professionals' educational attainment; and H3e: the integration strategy correlates positively with professionals' educational attainment), we used the previously used Kruskal-Wallis test.

Table 14- Kruskal-Wallis test for academic qualifications/conflict management

	Imposition			Accommodation			Commitment			Avoidance			Integration		
	Average	Standard deviation	Median	Average	Standard deviation	Median	Average	Standard deviation	Median	Average	Standard deviation	Median	Average	Standard deviation	Median
1st Elementary Cycle	2,4	0,9	2,5	3,9	0,9	4,0	4,0	0,8	3,7	3,9	1,2	4,0	4,0	0,8	4,0
2nd Basic Cycle	2,6	1,0	3,0	3,9	0,7	4,0	3,8	0,8	3,7	3,8	1,4	4,0	3,4	0,9	3,0
3rd Cycle Basic	2,6	1,0	3,0	3,9	0,7	4,0	3,9	0,7	4,0	4,0	0,9	4,0	3,8	0,7	4,0
Secondary Education	2,5	0,9	2,5	4,2	0,7	4,0	4,1	0,7	4,0	4,2	0,9	4,5	4,0	0,8	4,0
Bachelor's Degree	2,3	0,8	2,5	3,8	0,5	4,0	4,2	0,5	4,2	3,8	0,8	3,8	3,9	0,7	4,0
Degree	3,0	0,7	3,0	4,1	0,6	4,0	4,0	0,6	4,0	3,8	0,7	4,0	4,0	0,7	4,0
Master's Degree	3,0	0,7	3,0	3,8	0,0	3,8	3,8	1,2	3,8	4,3	1,1	4,3	3,8	0,0	3,8
Another	2,0	0,0	2,0	4,8	0,0	4,8	4,8	0,0	4,8	4,8	0,0	4,8	5,0	0,0	5,0
Chi-square	6,021			13,760			9,526			14,209			17,914		
Significance Sig.	,537			,056			,217			,048			,012		

After analyzing Table 14, we found that, in the imposition strategy, by determining the significance value, we found that the p value is higher than 0.05 (p=0.537), thus rejecting H3a, i.e. there are no significant differences between the imposition strategy and the level of education. With regard to the accommodative strategy, we found that the significance value is higher than 0.05 (p=0.056), thus we reject H3b, i.e. we assume that the level of education is not a determining factor in the accommodation variable. With regard to the commitment strategy, the significance level also reached values higher than 0.05 (p=0.217), thus we rejected H3c, i.e. academic qualifications do not influence the commitment variable. In the avoidance strategy, we found some oscillation between the means, from a minimum of 3.8 to a maximum of 4.8, confirming the acceptance of H3d, since the significance level gives values below 0.05 (p=0.048). This means that there are significant differences between the hospitality professionals' level of education and the avoidance variable. Finally, regarding integration, the averages reach a minimum value of 3.4 in the 2nd cycle and a maximum value of 5 in the "other" category. Also in the integration strategy, and based on the significance value p, we found a value of 0.012 (p<0.05), so we accepted H3e, i.e., the level of education is a determining factor in the integration strategy.

As regards the length of service and to test hypothesis 4 (H4a: the imposition strategy will be perceived differently depending on how long the employees have been working; H4b: the accommodation strategy will be perceived differently depending on how long the employees have been working; H4c: the commitment strategy will be perceived differently depending on how long the employees have been working; H4d: the avoidance strategy will be perceived differently depending on how long the employees have been working; H4e: the integration strategy will be perceived differently depending on how long the employees have been working), we used the Kruskal-Wallis test.

Table 15 - Kruskal-Wallis **test** for the time of professional exercise/management conflicts

| | Time of professional practice | | | | | | | | | Chi-square | Significance Sig. |
| | 1-15 | | | 16-25 | | | >25 | | | | |
	Average	Standard deviation	Median	Average	Standard deviation	Median	Average	Standard deviation	Median		
Imposition	2,8	1,2	3,0	3,0	1,3	3,3	2,3	0,9	2,5	1,967	,374
Accommodation	4,2	0,7	4,0	4,2	0,8	4,3	3,4	0,9	3,5	5,007	,082
Commitment	4,0	0,9	4,3	3,8	0,5	3,7	3,4	0,8	3,3	3,451	,178
Avoidance	4,0	1,1	4,3	3,9	1,0	3,8	3,6	0,9	3,5	1,215	,545
Integration	3,7	0,7	3,5	4,0	0,8	3,5	3,5	0,8	3,5	2,225	,329

Table 15 shows that the five conflict management strategies compared with the length of service variable have significance levels greater than 0.05, thus rejecting H4, i.e. there are no significant differences between the conflict strategies and the length of professional experience variable.

With regard to the employment relationship and in order to test hypothesis 5 (H5a: the imposition strategy correlates positively with the professionals' employment relationship; H5b: the accommodation strategy correlates positively with the professionals' employment relationship; H5c: the commitment strategy correlates positively with professionals' labor attachment; H5d: the avoidance strategy correlates positively with professionals' labor attachment; and H5e: the integration strategy correlates positively with professionals' labor attachment), we used the Kruskal-Wallis test.

Table 16 - Kruskall-Wallis test for employment bond/conflict management

	Employment relationship										
	Belongs to the Board			Contractor			Another			Chi-square	Significance Sig.
	Average	Standard deviation	Median	Average	Standard deviation	Median	Average	Standard deviation	Median		
Imposition	2,5	0,8	2,5	2,8	1,1	3,0	2,4	1,1	2,5	4,239	,120
Accommodation	4,1	0,7	4,0	4,0	0,8	4,0	4,0	0,6	4,0	2,982	,225
Commitment	4,1	0,7	4,0	3,8	0,8	3,7	4,1	0,6	4,0	9,683	,008
Avoidance	4,1	1,0	4,0	3,9	1,0	4,0	4,5	0,7	4,5	5,774	,056
Integration	4,0	0,8	4,0	3,8	0,8	4,0	4,3	0,9	4,5	8,373	,015

By analysing the values in Table 16, we found that both the imposition, accommodation and avoidance strategies assume significance levels higher than 0.05 (p=0.120 for imposition; p=0.225 for accommodation; p=0.056 for avoidance), thus rejecting hypotheses H5a, H5b and H5d. Therefore, we found that there are no significant differences between the employment relationship and the imposition, accommodation and avoidance strategies. However, we found that the variable employment relationship is a determining factor in the strategies of commitment and integration, thus we accepted H5c and H5e (p=0.08 in commitment; p=0.015 in integration).

As regards the training variable and for the testing of hypothesis 6 (H6a: the imposition strategy will be perceived differently depending on the training courses provided by the hotel; H6b: the accommodation strategy will be perceived differently depending on the training courses provided by the hotel; H6c: the commitment strategy will be perceived differently depending on the training courses provided by the hotel; H6d: the avoidance strategy will be perceived differently depending on the training courses provided by the hotel; H6e: the integration strategy will be perceived differently depending on the training courses provided by the hotel), we used the Mann-Whitney test.

Table 17- Mann-Whitney test for training/conflict management

| | Training courses provided | | | | | | | |
| | Yes | | | No | | | | |
	Average	Standard deviation	Median	Average	Standard deviation	Median	Z	Significance Sig. (2 extremities)
Imposition	2,6	0,9	2,5	2,5	0,9	2,5	-,893	,372
Accommodation	4,1	0,7	4,0	4,2	0,6	4,0	-,702	,483
Commitment	4,0	0,7	4,0	4,0	0,6	4,0	-,076	,939
Avoidance	4,0	0,9	4,0	4,0	1,0	4,5	-,069	,945
Integration	3,9	0,8	4,0	4,0	0,8	4,0	-,664	,507

Analyzing the significance values in Table 18, we find that all strategies have significance levels greater than 0.05 (p=0.372 in imposition; p=0.483 in accommodation; p=0.939 in commitment; p=0.945 in avoidance; p=0.507 in integration), therefore, we reject H6, i.e., the training variable does not influence the five conflict management strategies.

As regards the managerial position and for testing hypothesis 7 (H7a: the imposition strategy correlates positively with professionals in managerial positions; H7b: the accommodation strategy correlates positively with professionals in managerial positions; H7c: the commitment strategy correlates positively with professionals with management function; H7d: the avoidance strategy correlates positively with professionals with management function and H7e: the integration strategy correlates positively with professionals with management function) we used the Mann-Whitney test.

Table 18 - Mann-Whitney test for the leadership/conflict management role

| | Management or leadership role | | | | | | | |
| | Yes | | | No | | | | |
	Average	Standard deviation	Median	Average	Standard deviation	Median	Z	Significance Sig. (2 extremities)
Imposition	2,5	0,8	2,5	2,6	1,0	2,5	-1,115	,265
Accommodation	4,3	0,6	4,0	4,0	0,7	4,0	-2,895	,004
Commitment	4,2	0,6	4,3	3,9	0,7	4,0	-3,674	,000
Avoidance	4,0	1,0	4,0	4,0	1,0	4,0	-,016	,988
Integration	4,1	0,7	4,0	3,8	0,8	4,0	-3,431	,001

Examining the significance values in Table 18, we found that the imposition and avoidance strategies have values higher than 0.05 (p=0.094 in imposition; p=0.799 in avoidance), so we reject H7a and H7d. In other words, the leadership role does not influence the imposition and avoidance strategies. However, and according to the same table, we found that the strategies of accommodation, commitment and integration have significance values lower than 0.05 (p=0.04 in accommodation; p=0.000 in commitment; p=0.001 in integration). Thus, we accepted hypotheses H7b, H7c and H7e, which refer that the managerial role variable influences the strategies of accommodation, commitment and integration.

Next we test for normality and hypotheses regarding the motives of success, power, and affiliation.

In our case, and in relation to hospitality professionals in the R.A.M., it is found that the motives of affiliation, power and success do not approach the normal distribution, so we reject the null hypothesis (Ho: X has normal distribution) since our p<0.05. These results were obtained by applying the Kolmogorov-Smirnov normality test.

Table 19 - Kolmogorov-Smirnov normality test for the motivation variable

		Membership	Power	Success
N		349	349	349
Normal parametersa,b	Average	6,0823	2,8013	6,13
	Error Deviation	,81774	1,17121	,856
Most Extreme Differences	Absolute	,131	,077	,154
	Positive	,131	,077	,154
	Negative	-,131	-,062	-,142
Test statistics		,131	,077	,154
Significance Sig. (2 extremities)		,000c	,000c	,000c

a. The distribution of the test is Normal.

b. Calculated from data.

c. Lilliefors' correction for significance.

Having tested the normality of the motives of affiliation, power and success, we now check the various hypotheses of the study concerning the same motives.

The Mann-Whitney test was used to test hypothesis 8 (H8a: the success motive tends to be different according to the gender of the professionals; H8b: the affiliation motive tends to be different according to the gender of the professionals; H8c: the power motive tends to be different according to the gender of the professionals). The results obtained are presented in Table 20.

Table 20 - Mann-Whitney test for gender/motivation

	Sex of the participant							
	Female			Male				
	Average	Standard deviation	Median	Average	Standard deviation	Median	Z	Significance Sig. (2 extremities)
Success	6,2	0,8	6,4	6,0	0,9	6,2	-2,026	,043
Membership	6,1	0,8	6,3	6,0	0,9	6,1	-1,541	,123
Power	2,7	1,2	2,5	3,0	1,1	2,8	-3,070	,002

a. Grouping variable: Gender of the participant.

According to Table 20, we found that the motives of success and power when compared with the variable sex assume values lower than 0.05 (p=0.043 in success; p=0.002 in power), so we accept hypotheses H8a and H8c, i.e., sex influences the motives of success and power. On the other hand, the affiliation motive presents significance values (p=0.123) that show the inexistence of a relationship between the gender variable and this type of motive, so that we reject H8b. Considering the means for each gender, we found that women seem to be the ones who most often show up in the success motive. Contrarily, and in the power motive, we found that males seem to be the most frequent participants in this motive.

The Kruskal-Wallis test was used to test hypothesis 9 (H9a: The success motive will be perceived differently depending on the professionals' age; H9b: The affiliation motive will be perceived differently depending on the professionals' age; H9c: The power motive will be perceived differently depending on the professionals' age).

Table 21 - Kruskal-Wallis test for age/motivation

	Success			Membership			Power		
	Average	Standard deviation	Median	Average	Standard deviation	Median	Average	Standard deviation	Median
Up to 30 years old	5,9	1,1	6,0	5,9	1,0	6,1	2,8	1,3	2,5
30-39 years old	6,3	0,8	6,4	6,0	0,9	6,3	2,7	1,1	2,5
40-49 years old	6,1	0,8	6,2	6,2	0,7	6,3	2,8	1,2	2,7
50-59 years old	6,2	0,8	6,2	6,2	0,7	6,4	2,8	1,2	2,8
60-70 years old	5,9	0,9	5,8	5,9	0,8	6,1	3,2	1,4	3,3
Chi-square	5,71			3,21			2,27		
Significance Sig.	0,22			0,52			0,69		

a. Kruskal-Wallis test

b. Grouping variable: Age

According to Table 21, both the success motive and the affiliation and power motive show significance values greater than 0.05, so we reject H9a, H9b and H9c, i.e. there are no significant differences with respect to age for the three types of motives considered.

With regard to academic qualifications and to test hypothesis 10 (H10a: the success motive will be perceived differently depending on the academic qualifications of hospitality professionals; H10b: the affiliation motive will be perceived differently depending on the academic qualifications of hospitality professionals; H10c: the power motive will be perceived differently depending on the academic qualifications of hospitality professionals) we again used the Kruskal-Wallis test.

Table 22 - Kruskal-Wallis test for academic qualifications/motivation

	Academic qualifications								Statistics of testa,b	
	1st Cycle	2nd Cycle	3rd Cycle	Secondary Education	Bachelor's Degree	Degree	Master's Degree	Another	Chi-square	Sig.
	Average	Average	Average	Average	Average	Average	Average	Average		
Success	6,1	6,0	5,8	6,2	6,5	6,2	5,0	6,9	10,361	,169
Membership	6,0	5,8	6,0	6,2	6,1	5,9	5,4	6,6	10,755	,150
Power	2,8	2,4	3,0	2,8	2,6	3,0	2,5	2,4	4,442	,728

a. Kruskal-Wallis test.

b. Grouping variable: Educational qualifications.

According to Table 22, we found that the educational attainment does not influence the three motivational variables, since they assume significance values greater than 0.05, so we reject H10a, H10b and H10c.

In relation to the length of service and to test H11 (H11a: the success motive will tend to be different depending on the length of service of hospitality professionals; H11b: the affiliation motive will tend to be different depending on the length of service of hospitality professionals; the power motive will tend to be different depending on the length of service of hospitality professionals) we used once again the Kruskal-Wallis test.

Table 23 - Kruskal-Wallis test for the time of professional practice/motivation

	Time of professional practice									Statistics of test a,b	
	1-15			16-25			>25			Chi-square	Significance Sig.
	Average	Standard deviation	Median	Average	Standard deviation	Median	Average	Standard deviation	Median		
Success	6,2	1,1	6,4	5,8	1,0	5,9	5,0	1,3	5,2	5,806	,055
Membership	6,1	0,9	6,5	6,0	1,1	6,6	5,2	1,3	4,6	2,906	,234
Power	2,5	1,2	2,3	3,4	1,6	3,3	2,9	1,6	2,3	2,677	,262

a. Kruskal-Wallis test.

b. Grouping variable: Length of professional experience.

According to table 23 we found that the length of professional practice does not influence the motives of success, affiliation and power since the significance levels range between 0.055 and 0.262 (p>0.05).

With regard to the employment relationship and to test hypothesis 12 (H12a: the success motive will tend to be different depending on the organizational relationship of hospitality professionals; H12b: the affiliation motive will tend to be different depending on the organizational relationship of hospitality professionals; and H12c: the power motive will tend to be different depending on the organizational relationship of hospitality professionals) we used the Kruskal-Wallis non-parametric test.

Table 24 - Kruskal-Wallis test for employment bond/motivation

	Employment relationship									Test statistic [a,b]	
	Belongs to the Board			Contractor			Another				
	Average	Standard deviation	Median	Average	Standard deviation	Median	Average	Standard deviation	Median	Chi-square	Significance Sig.
Success	6,2	0,8	6,4	5,9	1,0	6,2	6,6	0,5	6,8	7,329	,026
Membership	6,2	0,7	6,3	5,8	1,0	6,0	6,5	0,6	6,7	9,774	,008
Power	2,8	1,2	2,7	2,6	1,1	2,3	3,0	1,5	3,3	2,907	,234

a. KruskalWallis test.

b. Grouping variable: Employment relationship.

According to the results presented in Table 24, the employment relationship of hospitality professionals influences both the success and affiliation motives (p=0.026 for success; p=0.08 for affiliation), which means that we accept H12a and H12b. Considering the means for each employment relationship, the category "others" seems to be the one that most often manifests itself with these types of motives (affiliation and success).

In what concerns training and to test hypothesis 13 (H13a: The success motive will tend to be different depending on the training opportunities of hospitality professionals; H13b: The affiliation motive will tend to be different depending on the training opportunities of hospitality professionals; H13c: The power motive will tend to be different depending on the training opportunities of hospitality professionals) we used the Mann-Whitney test. **Table 25** - Mann-Whitney test for training/motivation

	Training courses provided						Test statistic [a,b]	
	Yes			No			Z	Sig. (2 ends)
	Average	Standard deviation	Median	Average	Standard deviation	Median		
Success	6,1	0,9	6,4	6,1	0,8	6,2	-,457	,648
Membership	6,1	0,8	6,3	6,0	0,9	6,3	-,689	,491
Power	2,8	1,2	2,7	2,7	1,1	2,7	-,897	,370

a. Cluster variable: Training courses provided.

From table 25 we find that training courses provided by the organization do not influence the motives of success, affiliation and power (p=0.648 on success; 0.491 on affiliation; 0.370 on power), thus rejecting H13.

The Mann-Whitney test was used to test H14 (H14a: professionals with managerial positions tend to perceive the success motive differently; H14b: professionals with managerial positions tend to perceive the affiliation motive differently; H14c: professionals with managerial positions tend to perceive the power motive differently).

Table 26 - Mann-Whitney test for the leadership/motivation function

	Management or leadership role						Statistics	of the test [a]
	Yes			No			Z	Sig. (2 ends)
	Average	Standard deviation	Median	Average	Standard deviation	Median		
Success	6,3	0,7	6,4	6,1	0,9	6,2	-1,632	,103
Membership	6,1	0,7	6,3	6,1	0,9	6,3	-,125	,901
Power	2,9	1,1	2,8	2,8	1,2	2,5	-1,430	,153

a. Cluster variable: Managerial or leadership role

By looking at table 26 we confer that the significance levels vary between a minimum of 0.103 in success and 0.901 in affiliation. All values being greater than 0.05, we reject H14, i.e., the leadership role does not influence the three motives of McClelland (1987).

To test hypothesis 15 (hospitality professionals tend to positively correlate the strategies of imposition, accommodation, commitment, avoidance and integration with the motives of success, affiliation and power) Spearman's correlation was used. This correlation was introduced by Spearman (1904) and is defined as a statistic that necessarily requires two ordinal variables (X and Y). According to Maroco (2011), Pestana and Gageiro (2002), Spearman's correlation coefficient is a test used to measure the degree of intensity between ordinal variables and does not imply that the data collected come from populations whose normality is verified.

Table 27 - Spearman's correlation for conflict/motivation management

			Imposition	Accommodation	Commitment	Avoidance	Integration
Spearman's rô	Success	Correlation Coefficient	-,099	,373**	,458**	,299**	,390**
		Sig. (2 ends)	,066	,000	,000	,000	,000
		N	349	349	349	349	349
	Membership	Correlation Coefficient	-,160**	,405**	,425**	,321**	,305**
		Sig. (2 ends)	,003	,000	,000	,000	,000
		N	349	349	349	349	349
	Power	Correlation Coefficient	,277**	-,167**	-,093	-,173**	-,037
		Sig. (2 ends)	,000	,002	,082	,001	,490
		N	349	349	349	349	349

** Correlation is significant at the 0.01 level (2 ends).
* Correlation is significant at the 0.05 level (2 ends).

Table 27 shows that, with the exception of the imposition strategy, in which the correlation coefficient does not reach significant values, the success motive correlates positively with the strategies of accommodation (correlation coefficient = 0.373), commitment (correlation coefficient = 0.458), avoidance (correlation coefficient = 0.299) and integration (correlation coefficient = 0.390). It should also be noted that, with the exception of the imposition strategy (p=0.066), the levels of significance in the reason for success were always below 0.05, thus we concluded that the reason for success correlated positively with the strategies of accommodation (p=0.000), commitment (p=0.000), avoidance (p=0.000) and integration (p=0.00).

In the affiliative motive, we found a negative correlation between the affiliative motive and the imposition strategy (correlation coefficient = -0.160) and a positive correlation between the affiliative motive and the accommodation (correlation coefficient = 0.405), commitment (correlation coefficient = 0.425), avoidance (correlation coefficient = 0.321) and integration (correlation coefficient = 0.305) strategies. In addition, we mention that this reason when correlated with the conflict strategies always presents significance levels below 0.05 (imposition strategy: p=0.03;

accommodation: p=0.000; commitment: p=0.000; avoidance: p=0.000 and integration: p=0.00).

Finally, and with regard to the power motive, we found that this motive correlates positively with the imposition strategy (correlation coefficient = 0.277) and negatively with the accommodation (correlation coefficient = 0.167) and avoidance (correlation coefficient = 0.173) strategies. Furthermore, and to confirm the correlation of the imposition, accommodation and avoidance strategy with the power motive, we checked that the levels of significance in the three conflict strategies present values lower than 0.05 (imposition, p=0.00; accommodation, p=0.002 and avoidance, p=0.001). However, we found that the power motive does not correlate with the strategies of commitment and integration (commitment, p=0.082 and integration, p=0.490).

Chapter VII: Conclusions and Future Suggestions

7.1 Discussion of Results

The present research had as target audience the hospitality professionals of the R.A.M and has as main objective to relate conflict management with motivation, more precisely to understand the link between conflict management strategies (imposition, avoidance, compromise, accommodation and integration) and the three motives (power, success and affiliation) of David McClelland's theory.

Firstly, to investigate which is the predominant conflict management strategy in the hotel industry in the region, a questionnaire referenced by Jesuíno (2003 adapted from Thomas, 1976) was distributed to the professionals about the five conflict strategies previously applied in the Portuguese population. The results revealed that, in general, and on a scale of 1 to 5, individuals most frequently used the accommodation and avoidance strategy (mean accommodation = 4.09; mean avoidance = 4.04), and least frequently used the integration and imposition strategies (mean integration = 3.92; mean imposition = 2.55).

In both accommodation and avoidance strategies, hotel professionals showed little concern in satisfying their own interests. Also in Rego's study (1995), the accommodation strategy reached one of the highest averages when compared to the other conflict strategies. In turn, Popescu (2015), when relating the three attitudes towards change (the personal factor, the rational factor and the conservative factor) with the five conflict strategies in his study, found that the conservative factor had a higher percentage in the avoidance strategy and a lower percentage in the collaboration or integration strategy. McIntyre (2007) explains that individuals more often use the avoidance and accommodation strategies because they are not very receptive to dialogue, thus contributing to a reduced interaction and participation in the organisational environment. This fact can be improved by improving the quality of communication through the development of speaking and listening skills (Kiralp, Dincyürek, & Beidoglu 2009). According to the same authors, what often happens is that individuals are more concerned with defending their theory (through argumentation) than properly listening to the other's opinion.

Secondly, and in order to analyse which type of motive is more present in the hotel industry of the region, we used Rego and Carvalho's (2001, 2002) questionnaire regarding success, affiliation and power motives already applied in previous studies and to the Portuguese population. (Rego & Leite, 2003; Rego, Tavares, Cunha & Cardoso, 2005).

The results of this study revealed that on a scale of 1 to 7, individuals seem to be more motivated for the success motive (mean=6.13) and affiliation (mean=6.08), denoting a low trend for the power motive (mean=2.80). Similarly, the study of Rego (2000) found that managers within the organization showed to be more motivated for success denoting a mean of 5.6 and less for affiliation (mean of affiliation=5.5) and power (mean of power=4.0). Also the study of Rego, Tavares, Cunha and Cardoso (2005) revealed that individuals tended to be more motivated for the success motive than for affiliation and power. Contrarily, in the study by Rego (1998) individuals were more motivated towards affiliation (mean=5.7) and less motivated towards success (mean=4.7) and power (mean=4.1).

Both motivation and conflict management depend on socio-demographic aspects which include personal and professional characteristics. Therefore, we first present the discussion of the results of the hypothesis tests related to conflict management strategies (imposition, accommodation, compromise, avoidance, and integration) and, subsequently, the discussion of the results of the tests related to motivation (power motive, success, and affiliation).

The results of the present study revealed that the variable gender is not a determining factor for conflict management strategies (H1 rejected). Nicotera and Dorsey (2006, p.312) confirm this fact, stating that "conflict styles are not driven by gender (...) the search for gender differences in organisational communication and conflict particularly have little expectation of producing any significant results". Also Wall and Blum (1991) refer that there is a weak and inconsistent relationship between the variable gender and conflict management strategies.

With regard to the variable age and its relationship with conflict strategies, and taking into account the results of this study, we found that, in general, the variable age shows a weak relationship with conflict strategies, except for the strategy of imposition, which shows significant differences in this variable (H2a accepted; H2b; H2c; H2d and

H2e rejected). Morrow and Wirth (1989) also mentioned that there is a weak relationship between the variable age and conflict management strategies.

Marques and Cunha (2004), when analysing the determinants of conflict management, found that individuals with secondary and higher education more often used imposition (or domination) and avoidance strategies. In contrast, Munduate, Ganaza, and Alcaide (1993) found that more educated individuals, when faced with conflict situations, more often use integration and avoidance strategies. This study also revealed significant differences between the hotel professionals' academic qualifications and the integration and avoidance strategies (H3a rejected; H3b rejected; H3c rejected; H3d accepted; H3e accepted).

No significant differences were found regarding the length of professional experience and conflict management strategies (H4 rejected). However, Marques and Cunha (2004) revealed in their study that individuals with higher scores in the imposition strategy (domination) had between 0 and 15 years of service or more than 30 years of service. When investigating the impact of the length of professional experience on negotiation effectiveness, Cunha (2003) also found that this correlated positively with the length of service variable.

The results also revealed that the employment relationship of hospitality professionals in the A.R. only influences commitment and integration strategies, and that there are no significant differences between the employment relationship and imposition, accommodation and avoidance strategies (H5c and H5e accepted; H5a, H5b and H5d rejected). The study conducted by Misquita (referred to by Rahim, 2001) also revealed the existence of a positive correlation between organizational commitment and the integrative strategy, in which superiors, by managing conflict in an integrative way, encouraged the professional to commit more to the organization. On the contrary, if superiors chose an avoidance or imposition strategy, the subordinate's commitment to the organisation decreased (Rahim, 2001).

In this study, no significant differences were found between conflict management strategies and the variable training (H6 rejected). However, McIntyre (2007) states that managers must have the ability to manage conflict effectively by using certain strategies necessary to resolve conflict, and it is also important for them to be aware of their difficulties and limitations, seeking training in the area of conflict management. According to the same author, training is an indispensable element to the organisation, since through training actions provided to managers and subordinates, it is possible to increase competitive advantage and organisational profitability.

As for the impact of the managerial position on conflict management, we found that the managerial position held by hospitality professionals was a determining factor for the use of the strategies of accommodation, commitment and integration, and did not influence the strategies of imposition and avoidance (H7a and H7d accepted; H7b, H7c and H7e rejected). Contrarily, Jesuíno (1992) revealed in his study that individuals with management positions within the organisation were more motivated to succeed and tended to use the imposition strategy more often. On the contrary, individuals with low organisational power chose to use the integration and avoidance strategy.

Next, we present the results concerning the hypotheses that relate the sociodemographic variables and the motives of success, power and affiliation.

According to our study, we can conclude that the variable gender of hotel professionals influences the motives of success and power, unlike the motive of affiliation (H8a and H8c accepted; H8b rejected). Since gender is an influencing variable of the success and power motives, we also found that, while females seem to be more motivated towards the success motive, males seem to be more motivated towards the power motive. These results are in line with the study conducted by Rego and Jesuíno (2002) who revealed that males tend to use the power motive more often, unlike females. Rego (2000) did not find any significant differences between the gender variable and the power motive and success however, female individuals seemed to be slightly more motivated towards affiliation when compared to males. Kukanja (2013), when comparing the sex variable with the motivating factors of money, career development, training, security, fun and flexible schedule, found that sex influences money and fun, with money and fun being more important to female than male individuals. To some extent, we can point to cultural and socialisation differences between men and women that persist to the present day as hypotheses for this difference in results (Amâncio, 2001).

When comparing the variable age with the professionals' motives of success, affiliation and power, we found that these motives do not present any significant differences with the variable age, i.e. the hospitality professionals' age does not influence the fact that individuals are more oriented towards the motives of success, affiliation or power (H9 rejected). However, Luthans and Thomas (referred to by Cunha et al, 2007) found that age influences satisfaction. These same authors conferred that the most satisfied individuals are the older ones, since they have a greater possibility of reaching leadership positions that assign more responsibilities and enable higher salaries, making the reality of professional life meet the employees' expectations.

On the other hand, the level of education of hospitality professionals in this study does not influence the motives of power, success and affiliation (H10 rejected). However, according to the study carried out by Rego (2000), education is inversely correlated with the affiliative motive, i.e. individuals with less education tend to be more affiliative and vice-versa. In turn, Rego, Tavares, Cunha, and Cardoso (2005) argue that students who aspire to achieve leadership and prestigious positions seem to be more motivated towards success and power than those with lower levels of education. This is explained by the fact that these individuals, unlike in the past, usually invest their own money and often have to reconcile their personal and professional lives. The fact that they have to endure such efforts and often give up their personal and professional lives makes them more competitive and ambitious than less educated individuals. Fabra and Camisón (2009) conclude that individuals with higher education and bachelor's degrees are less satisfied than individuals with an education equivalent to basic education.

Also from our study, we found that the length of professional experience does not influence the professionals' reasons for success, power and affiliation (H11 rejected). Also Ronen (referred to by Romão, 2011) revealed that the length of service does not influence work motivation, since individuals are less satisfied at work in their first years, remaining so for some time. However, after this period, the individuals' satisfaction grows again, increasing, as a result, the worker's expectations (Ronen, referred to by Romão, 2011). However, if these expectations are frustrated, the level of satisfaction returns to zero.

When comparing the employment relationship variable with the motives of success, affiliation and power of hospitality professionals, we found that this variable shows significant differences regarding the motives of success and affiliation (accepted hypotheses H12a and H12b). Tezergil, Köse, and Karabay (2014) argue that individuals with high levels of self-esteem, motivation, and seniority are highly committed to their work. According to the same authors, one of the consequences of high levels of motivation and work engagement is the high *performance of* professionals, since when professionals are motivated and engaged, they are more committed to their workplace.

No significant differences were found between training and the motives of success, affiliation and power (H13 rejected). According to Lacerda and Abbad (2003) for training to influence motivation it is important that trainees know and perceive the importance of the "instructional" programmes (profits achieved from them). Contrarily and according to the same authors, programs that do not involve future projects do not allow the achievement of the desired goals, leading to demotivation. For the managers of the organisation to meet the needs of their trainees, it is important to understand their needs, creating for example, continuous training programmes so that the trainees permanently receive academic and professional training, important for their professional activity.

As regards the managerial position and its relationship with the motives of success, power and affiliation of hospitality professionals, no significant differences were found (H14 rejected). On the contrary, studies by Rego and Jesuíno (2002) and Rego (2000) found that individuals with a degree in management had a greater tendency towards the success and power motive than the affiliative motive.

Finally, we proceed to discuss the central hypothesis (H15) of this paper which states that hospitality professionals tend to positively correlate the strategies of imposition, accommodation, compromise, avoidance and integration with the motives of success, affiliation and power.

The present study revealed that the need for success correlates with the strategies of accommodation, compromise, avoidance and integration, while no significant differences were found in the strategy of imposition. Similarly, the study conducted by Bell and Blakeney (1977) revealed that the need for success correlates positively with integration (need to confront).

As regards the affiliative motive, and in accordance with our study, we found that this motive correlated negatively with the imposition strategy and positively with the accommodation, commitment, avoidance and integration strategies. Similarly, the study by Schneer and Chanin (1987) revealed that the need for affiliation correlates positively with the accommodation strategy and negatively with the imposition strategy. It is hypothesised that affiliation correlates negatively with imposition because individuals with greater social and affective skills, when facing a conflict situation, avoid using the competitive environment as a way of solving their problems, since one of the goals of affiliative individuals is to maintain the affective ties and the cohesion of the work group.

Finally, the results revealed that the power motive correlates positively with the imposition strategy and negatively with the accommodation and avoidance strategy, while no relationship was found between the power motive and the commitment and integration strategies. Also Bell and Blakeney (1977) stated that individuals who were more motivated towards power used the imposition strategy as a way to solve their conflicts. In this sense, and according to the same authors, the power motive correlates positively with strategies involving the obligation to force the other to agree with their opinion (imposition strategy) and negatively with strategies involving lightness in conflict resolution (accommodation and avoidance strategies). In some way, we can point to the fact that individuals who like to influence others use all possible means to achieve the desired goals, imploring and forcing the other to satisfy their needs as explanatory hypotheses for this relationship between the power motive and the imposition strategy.

7.2 Final considerations

The main objective of this work was to relate conflict management strategies with the motivations of hospitality professionals. To this end, we resorted to the strategies of avoidance, imposition, integration, accommodation and commitment and to David McClelland's theory which distinguishes three distinct types of motives: the motives of success, affiliation and power.

The study was based on the quantitative methodology. Thus, and in order to better understand which conflict management strategies and motives predominate in the hotel industry in the Madeira Region we used the questionnaire referenced by Jesuíno (2003 adapted from Thomas, 1976) and the questionnaire of Rego and Carvalho (2001, 2002), already previously validated.

The study in question thus contributed to a better understanding of this organizational reality (conflict and motivation) of hospitality professionals who work in the hotel industry in Madeira. The results show that not all conflict strategies (avoidance strategy, imposition, integration, commitment and accommodation) correlate with all hotel professionals' motivations (success, power and affiliation motives). More specifically, we found that the success motive correlates positively with the strategies of accommodation, commitment, avoidance and integration, revealing no relationship with the imposition strategy. The affiliation motive correlates negatively with the imposition strategy and positively with the accommodation, commitment, avoidance and integration strategies. As regards the power motive, we found that it correlates positively with the imposition strategy and negatively with the accommodation and avoidance strategies, revealing the lack of relationship with the commitment and integration strategies.

Also when analysing the conflict management strategies, we found that accommodation had a higher mean score than the other strategies. This may be explained by the fact that people feel uncomfortable participating, thus limiting their intervention within the organisation. Thus, in order for professionals to have a more active voice in the organisation, it is important to improve the communication between organisational members. Improving internal communication allows clarifying misunderstandings and possible doubts that may arise on certain issues. To this end, it is suggested that the hotel human resources management be concerned with organising more work meetings.

The increase in participatory management within the hotel units would allow for an increase in job satisfaction (Kim, Spence-Laschinger & Finegan referred by Shagholi, Abdolmalki & Moayedi, 2011), a better achievement of objectives, a better and efficient decision-making (Hargreaves & Hopkins, Likert referred by Shagholi, Abdolmalki & Moayedi, 2011), greater social responsibility within the organization (Eisenberger et al referred to by Shagholi, Abdolmalki & Moayedi, 2011), better people management (Ospina & Yaroni referred to by Shagholi, Abdolmalki & Moayedi, 2011) and consequent conflict management and finally, an increase in organizational *performance* (Lau Lim & Ming referred to by Shagholi, Abdolmalki & Moayedi, 2011). The increase in participatory management would also allow for greater information sharing achieved through greater communicational openness and more permissive authority and greater power autonomy in the workplace (Wood et al. referred to by Shagholi, Abdolmalki, & Moayedi, 2011).

In addition to these conclusions, we were able to draw other conclusions regarding the socio-demographic variables and their association with conflict management. We concluded that the variable gender does not influence conflict management strategies; however, we can say that the variable age influences the imposition strategy. In relation to academic qualifications, we found that these influence the strategies of avoidance and integration used by hospitality professionals. In what concerns the length of professional experience and the professionals' training, we found that these are not a determining factor for the avoidance, accommodation, imposition, integration and commitment strategies. Significant differences were also found between the employment relationship and the commitment and integration strategies. Finally, we confirmed that the leadership role influences the professionals' accommodation, commitment and integration strategies.

In this study, we also tried to understand the relationship between sociodemographic variables and the motives for success, power and affiliation. We found that the gender variable influences the motives for success and power of hospitality professionals, with no significant differences between the gender variable and the affiliation motive. With regard to the variables age, academic qualifications, length of professional experience, training and leadership role, we found that these variables do not influence the motives for success, affiliation and power. In relation to the employment relationship, the study revealed that this is a determining factor for hospitality professionals who seem to be more motivated towards success and affiliation.

This study contributed to deepen and understand the importance of a good conflict management and the motivation of hospitality professionals. In this sense, we tried to convey some basic knowledge about the conflict strategies that are important for the effectiveness of problem management in an organisational context, as well as to understand the professionals' main motivations. The present study also presents some limitations that may condition the results.

One of the limitations of the study is the fact that this research cannot be generalized to other cases because it is a very specific theme, but also because it is a convenience sample in which respondents are chosen according to their availability and adherence. Another difficulty encountered in the research was the lack of adherence by some hotel establishments which hampered data collection, requiring the expenditure of more time to contact other possible establishments. As previously mentioned, the sample is mainly composed of individuals who do not hold managerial or leadership positions, which constitutes an obstacle for the research, since one of the objectives of the study was to study the impact of the managerial role on conflict management and motivation. For a better understanding of this phenomenon, we could use another type of research methodology, such as the interview.

As for suggestions for future research, we believe that it would be useful, as previously mentioned, to focus on a study which is not only quantitative but also qualitative, allowing us to answer all the questions related to the topic. Another suggestion would be to study to what extent employees' personalities affect the relationship between conflict management and motivation. Another suggestion would be to relate the theme of conflict management to another theory of process motivation, such as Vroom's expectations theory.

It is also recommended that studies on conflict management strategies and motivation be further explored by organizations in order to minimize the negative effects of conflict and allow for a more stable and calm environment, conducive to the motivation of professionals. For this, organizations in general should invest more in the training of their professionals in the area of conflict management, increasing their social and affective skills.

As the hospitality industry is a very broad area, it would also be interesting for future research purposes to compare this study with similar studies but with professionals who hold only managerial or prestigious positions. The focus on professionals with management positions would allow understanding how superiors deal with the conflict and understanding their main motivations.

It is also suggested for future research to study the theme of conflict and motivation in hotel units belonging only to the four-star category that holds a more personalized and qualified service than hotels belonging to the categories of one, two and three stars.

It would also be enriching to analyze the correlation of other sociodemographic characteristics with conflict and motivation strategies, for example marital status and workload.

Since the hotel business is a very wide area, it would also be important to compare the present study, made only from hotel establishments belonging to the category of hotel with establishments belonging to the categories of boarding houses, inns and tourist apartments.

In relation to the data collection instruments used, they could still be improved through factor analysis, so as to better visualise the data and make it easier to interpret them.

In conclusion, the importance of a continuous study on this topic is reinforced. All institutions should assume the importance of conflict management and motivation of their professionals. Although some institutions associate the term conflict with something harmful, conflict is something inevitable in an organisation. Thus, it is important to know how to manage it in order to stabilize and create an environment conducive to the harmony and motivation of its professionals.

Bibliographic References:

Alves, E., Dacoreggio, M., Becker, F. & Teixeira, G. (2008). *Metodologia: construção de uma proposta científica* (Ed. 1ª). Curitiba: Editora Camões.

Amâncio, L. (2001). Gender in psychology: a history of mismatches and ruptures, *Psicologia*, Vol. XV (1), pp. 9-26.

Antonioni, D. (1998). Relationship between the big five personality factors and conflict management styles. *International Journal of Conflict Management*, 9 (4), 336-355.

Barañano, A. (2004). *Métodos e técnicas de investigação em gestão: manual de apoio à realização de trabalhos de investigação* (Ed. 1ª). Lisboa: Sílabo.

Bell, E. & Blakeney, R. (1977). Personality correlates of conflict resolution modes. *Human relations*, 30, 849-857.

Bilhim, J. (1996). *Teoria organizacional*. Lisboa: Instituto Superior de Ciências Sociais e Políticas.

Bipp, T. & Dam, K. V. (2014). Extending hierarchical achievement motivation models: The role of motivational needs for achievement goals and academic performance. *Personality and Individual Differences*, 64, 157-162.

Bruce, A. & Pepitone, J. (2002). *Motivating employees*. Lisboa: McGraw-Hill.

Bussad, W. & Morettin, P. (2002*). Estatística Básica* (Ed. 5ª). São Paulo: Saraiva

Caetano, A. & Vala, J. (2000). *Gestão de Recursos Humanos : contextos, processos e técnica*. Lisboa: RH Editora.

Camilleri, E. (2007). Antecedents affecting public service motivation. *Personnel Review*. Emerald Group Publishing Limited, 36 (3), 356-377.

Castelli, G. (1992). *Administração hotelira* (3ª Ed.). Caxias do Sul: Editora Universidade de Caxias do Sul. P- 37.

Cavalcanti, A. (2006). *O Gestor e o seu papel na Gestão de Conflitos: um estudo de caso em empresa Varejo de Vestuário masculino.* Tese de Mestrado em Administração e Pesquisas de Administração. Belo Horizonte: Federal University of Minas Gerais.

Chiavenato, I. (1987). *Administração RH.* São Paulo: Atlas.

Chiavenato, I. (1998). *Recursos Humanos* (Ed. 5ª). São Paulo: Atlas.

Chiavenato, I. (1999). *Gestão de pessoas: o novo papel dos recursos humanos nas organizações.* Rio de Janeiro: Campus

Chinyowa, K. (2013). Exploring Conflict-Management Strategies through Applied Drama: A Wits University Case Study. In Barnes, H. (ed.). *Arts Activism, Education and Therapies: Transforming Communities across Africa,* 44, 39 - 54.

Chrispino, A. (2007). *School conflict management: from conflict classification to mediation models. Ensaio: aval. pol. publ. Educ,* 15 (54), 11-28.

Cortina, J. M. (1993). What is coefficient alpha? An examination of theory and applications. *Journal of Applied Psychology,* 78, 98-104.

Coutinho, P. (2008). Correlational studies in education: potentialities and limitations. *Psicologia Educação e Cultura,* XII (1), 143-169.

Cunha, P. (2001). *Conflito e Negociação* (Ed.1ª). Porto: Editoras Asa.

Cunha (2003). Algumas reflexões sobre eficácia em negociação: resultados de um estudo experimental. *IV Congresso Português de Sociologia.*

Cunha, M., Rego, A., Cunha R. & Cardoso, C. (2007). *Manual de Comportamento Organizacional e Gestão.* Lisboa: Editora RH.

Cunha, P. (2008). *Conflito e Negociação* (Ed.1ª). Porto: Editoras Asa.

Deci, E. L. & Ryan, R. (1985). *Intrinsic Motivation and Self-Determination in Human Behavior.* New York: Plenum Press.

Dessler, G. (2004). *Human Resource Management* (Ed. 10ª). Prentice Hall

Dickson, R., Ford, R. C. & Upchurch, R. (2005). A case study in hotel organizational alignment. *The International Journal of Hospitality Management,* 25, 463-477.

Dimas, I., Lourenço, P. & Miguez, J. (2005). Conflitos e Desenvolvimento nos Grupos e Equipas de Trabalho - uma abordagem integrada. *National Scientific Journal,* 38 (1), 103-119.

Dimas, I. D., Lourenço, P. L. & Miguez, J. (2007). (RE)thinking about intragroup conflicts: performance and developmental levels. *Psychology,* 21 (2), 183-205.

Regional Directorate of Statistics (2014). "Statistical indicators." On the website: http://estatistica.gov-madeira.pt, on November 5, 2014.

Dune, M. J. (1989). Sex differences in styles of conflict management. *Psychological Reports,* 65, 1033-1034.

Enz, Cathy A. (2001). What Keeps You Up at Night? Key Issues of Concern for Lodging Managers. *The Cornell Hotel and Restaurant Administration Quarterly,* 42(2), 38-45.

Fabra, M. & Camisón, C. (2009). Direct and indirect effects of education on job satisfaction: A structural equation model for the Spanish case. *Economics of Education Review,* 28(5), 600-610.

Ferreira, J., Neves, J. & Caetano, A. (2001). *Manual de Psicossociologia das Organizações.* Lisboa: Escolar Editora.

Ferreira, J., Neves, J. & Caetano, A. (2011). *Manual de Psicossociologia das Organizações.* Lisboa: Escolar Editora.

Filho, J. P., Viola, D. N. & Borges, G. (2010). *Using the randomization test to compare two groups considering non-parametric test. National Symposium.* Departamento de Estatística, Universidade Federal da Bahia.

Fortin, M. (2000). *O processo de investigação: Da conceção à realização* (Ed. 2ª). Loures: Lusociência.

França, C. B. & Lourenço, P. R. (2010). Diversity and intragroup conflict at work: an empirical study in Portugal. *Revista de Administração Mackenzie,* 11 (3), 130-158.

Freeman, B. (1994). Power motivation and youth: An analysis of troubled students and student leaders. *Journal of Counseling and Development,* 72 (6), p.661-671.

Freire, J. (1942). *Sociologia do trabalho: uma introdução* (Ed.2º). Porto: Edições afrontamento.

Freixo, M. (2011). *Metodologia Científica: Fundamentos, Métodos e Técnicas* (Ed. 3ª). Lisboa: Epistemologia e Sociedade.

Galhanas, Carla R. G. (2009). *A motivação dos recursos humanos nos novos modelos de gestão da administração pública.* Tese de Mestrado em Ciências Empresarias. Lisboa: Instituto Superior de Economia e Gestão.

Gil, A. (2008). *Métodos e técnicas de pesquisa social* (6ª Ed.). São Paulo: Atlas.

Goel, D. (2012). Exploring the Predictive Power of Demographic Factors on Conflict Management Styles of Individuals: A study of Moserbaer Photovoltaic Ltd. *Drishtikon Management Journal,* 3 (1), 76-97.

Gottfried, A. E., Fleming, J. M. & Gottfried, A. W. (2001). Continuity of academic intrinsic motivation from childhood through late adolescence: a longitudinal study. *Journal of Educational Psychology,* 93 (1), 3-13.

Hayes, B. E. (1998). *Measuring Customer Satisfaction: Survey design, use, and statistical analysis methods.* Milwaukee, Wisconsin: ASQC Quality Press.

National Statistical Institute (2014[a]). *"Employment Statistics".* On the website: ine.pt.

Instituto Nacional de Estatística (2014 [b]). *Hotel Establishments, Holiday Villas and Apartments, by NUTS II and Typology.* On the website: http://www.turismodeportugal.pt, on 15 November 2014.

Iochida L. C. & Castro A. A. (2001). Research design (Part VIII - statistical method / statistical analysis). In: Castro AA. *Planejamento da Pesquisa.* São Paulo: ACC.

Jesuíno (1992). *A negociação- Estratégias e Táticas.* Lisboa: Texto Editora.

Jesuíno, J. C. (2003). *A negociação: Estratégias e Táticas.* Lisboa, Texto Editora.

Jesus, S. N. (2000). Motivação e formação de professores. *Revista Lusófona de Educação,* p.205-208.

Karakus, M. & Savas, A. C. (2012). The Effects of Parental Involvement, Trust in Parents, Trust In Students and Pupil Control Ideology on Conflict Management Strategies of Early Childhood Teachers. *Educational Sciences: Theory and Practice,* 2977-2985.

Kaushal, R. & Kwantes, C. T. (2006). The role of culture and personality in choice of conflict management strategy. *International Journal of Intercultural Relations, 30,* 579-603.

Kiralp Y., S. Dincyürek & M. Beidoglu (2009). Determining the conflict resolution strategies of university students. *Procedia-Social and Behavioral,* 1 (1), 1205-1214.

Koenes, A. (1996). *Gestión y motivación del personal.* España: Ediciones Díaz de Santos.

Kolb, D. & Putman, L. (1992). The multiple faces of conflict in organizations. *Journal of Organizational Behaviour.*

Kukanja, M. (2013). Influence of demographic characteristics on employee motivation in catering companies. *Tourism and Hospitality Management,* 19 (1), 97-107.

Lacerda, E. & Abbad (2003). Impacto do Treinamento no Trabalho: Investigando Variáveis Motivacionais como suas Preditoras. *Journal of Contemporary Administration,* 7 (4), 77-96.

Lobos, J. A. (1978). *Comportamento Organizacional.* São Paulo: Atlas.

Loureiro, L. & Gameiro, M. (2011). Interpretação crítica dos resultados estatísticos: para além da significância estatística. *Revista de Enfermagem Referência,* 3 (3), 151-162.

Marconi, M & Lakatos, E. (2003). *Fundamentos de metodologia científica* (5ªEd.). São Paulo: Atlas.

Maroco, J. & Garcia-Marques, T. (2006). What is the reliability of Cronbach's alpha? Old questions and modern solutions? *Laboratório de Psicologia,* 4 (1), 65-90.

Marôco (2011). *Statistical Analysis with SPSS Statistics* (5th Ed.). Pero Pinheiro: Report Number

Marques, L. & Cunha, P. (2004). Estilos de gestão de conflitos em contexto escolar: Análise de algumas variáveis relevantes. *Proceedings of the workshops of the V° Congresso Português de Sociologia - Sociedades Contemporâneas - Reflexividade e Acção.*

McAdams, D. P. (1992). The five-factor model in personality: A critical appraisal. *Journal of Personality, 60,* 329-361.

McClelland, D. C. (1965). Achievement and entrepreneurship: A longitudinal study. *Journal of Personality and Social Psychology,* 1, 389-392.

McClelland, D. C. (1987). *Human Motivation* (2ª Ed.). Cambridge: Cambridge University Press.

McIntyre, S. (2007). How people manage conflict in organizations: Individual negotiating strategies. *Psychological Analysis*, 2 (xxv), 295-305.

Miner, J. B. (2005). *Organizational Behaviour: essential theories of motivation and leadership.* Oxon and New York: Routledge.

Morberg, P. J. (2001). Linking conflict strategy to the five-factor model: Theoretical and empirical foundations. *International Journal of Conflict Management,* 12(1), 47-68.

Morrow, P. C. and Wirth, R. E. (1989). Work commitment among salaried professionals. *Journal of Vocational Behavior,* 34, 40-56.

Moscovici, F. (2001). *Desenvolvimento Interpessoal: Treinamento em grupo* (Ed. 10ª). Rio de Janeiro: José Olympio.

Muchinsky, P. M. (2004). *Psicologia Organizacional.* São Paulo: Pioneira Thomson Learning.

Munduate, L, Ganaza, J & Alcaide, M. (1993). Estilos de géstion del conflicto interpersonal en las organizaciones. *Revista de Psicologia Social,* 8(1), 47-68.

Murray, H. A. (1938). *Explorations in personality.* New York: Oxford University Press.

Nath, R. & Raheja, R. (2001). Competencies in hospitality industry, *Journal of Services Research,*

1(1), 25-33.

Neves, A. (1998). *Motivação para o trabalho* (Ed. 1ª). Lisboa: RH Editora.

Nickson, D. (2007). *Human resource management for the hospitality and Tourism Industries.* Oxford: Elsevier.

Nicotera, A. & Dorsey, L. (2006). Individual and interactive processes in organizational conflict. In Oetzel.J & Ting-Toomey. *The SAGE handbook of conflict communication: Integrating Theory, Research, and Practice. P.* 293-327

Noltemeyer, A., Bush, K., Patton, J. & Bergen, D. (2012). The relationship among deficiency needs and growth needs: An empirical investigation of Maslow's theory. *Children and Youth Services Review,* 34, 1862-1867.

Perez-Ramos, J. (1990). Motivation at work: theoretical approaches. *Psicologia-Universidade de São Paulo,* 1 (2), 127-140.

Pestana, M. & Gageiro, J. (2002). *Análise de Dados para Ciências Sociais: A complementaridade do SPSS.* Lisboa: Edições Sílabo, 2002.

Popescu, A. M. (2015). Prescriptive Models of Intervention Strategy Choice of Manager in the Resolution of Conflict Moods, *Procedia-Social and Behavioral Sciences*, 180, 197-202.

Pruitt, D. G. & Rubin, J. Z (1986). *Social conflict, Escalation, Stalemate, and Settlement*. New York: Random House.

Quivy, R. & Campenhoudt, L., V. (2005). *Manual de investigação em ciências sociais*. Lisboa: Gradiva.

Rahim, M. (2001). *Managing Conflict in Organizations* (3ª Ed.). London: Quorum Books.

Rego (1995). *O modelo Motivacional de McClelland - uma aplicação*. Master's Dissertation in Business Science. Lisboa: ISCTE

Rego, A. (2000). The motives of success, affiliation and power - development and validation of a measurement instrument. *Psychological analysis, 3* (XVIII), 335-344.

Rego, A. & Carvalho, T. (2001). The achievement, affiliation and power motives: a confirmatory study of the three factor-model. *Journal of Applied Psychology (Angewandte Psychologie)*, 3 (4), 89-104.

Rego, A. & Carvalho T. (2002). The motives of success, affiliation and confirmatory construct-evidence. *Psychology: Theory and Research*, 18 (1), 17-26.

Rego, A. & Jesuíno, J. (2002). Comportamento *organizacional e gestão*. Lisboa: Instituto superior de Psicologia Aplacada, 8 (1): 83-97.

Rego, A. & Leite, E. (2003). Motives for success, affiliation and power: a construct validation study in Brazil. *Psychology Studies* 8 (1), 185-191

Rego, A., Tavares, A. I., Cunha, M. P. & Cardoso, C. C. (2005). The motives of success, affiliation and power: undergraduate and graduate students' motivational profiles and their relationship with remuneration levels. *Psicologia reflexão e crítica,* 18 (2), 225-236.

Renwick, P. A. (1975). Perception and management of superior-subordinate conflict. *Organizational Behavior and Human Performance, 13*, 444-456.

Robbins, P. (2002). *Comportamento Organizacional.* São Paulo: Prentice Hall.

Robinson, J. (2000). *Mastering Motivation.* Queluz de Baixo: Expansion Books.

Rocha, J. (2010). *Gestão de Recursos Humanos na Administração Pública.* Lisboa: Escolar Editora.

Romão, P. C. (2011). A *satisfação laboral e a perceção da qualidade dos serviços prestados: estudo de caso de uma autarquia.* Dissertação de Mestrado em Gestão Estratégica de Recursos Humanos, Instituto Politécnico de Setúbal.

Schermerhorn, J. R., Hunt, J. G., & Osborn, R. N. (1999). *Fundamentos de comportamento organizacional.* Brazil: Bookman.

Schneer, J. A., & Chanin, M. N. (1987). Manifest needs as personality predispositions to conflict-handling behavior. *Human Relations*, 40(9), 575-590.

Shagholi, Abdolmalki & Abdolmalki (2011). New Approach in Participatory Management, Concepts and Applications, *Procedia Social and Behavioral Sciences,* 15, 251-255.

Simpson, C. (1998). *Coping through conflict resolution and peer mediation.* New York: Skills Library.

Spearman, C. (1904). General intelligence, objectively determined and measured. *American Journal of Psychology, 15,* 201-293.

Spinelli, M. A. & Canavos, G.C.. (2000). Investigating the Relationship between Employee Satisfaction and Guest Satisfaction. *The Cornell Hotel and Restaurant Administration Quarterly,* 41(6) 29-33.

Stoner, J. A. F. & Freeman, R. E. (1999). *Administração.* Rio de Janeiro: LTC.

Streiner, D. L. (2003). Being inconsistent about consistency: when coefficient alpha does and doesn't matter. *Journal of Personality Assessment,* 80, 217-222.

Tamayo, A. & Paschoal, T. (2003). The Relationship of Work Motivation to Work Goals of the Worker. *Journal of Contemporary Administration,* 7 (4), 33-54.

Taraboulsi, F. (2003). *Administração de hotelaria hospitalar.* São Paulo: Atlas Publishing House.

Tezergil, S. A., Köse, A. & Karabay, M. E. (2014). Investigating the Effect of Trust, Work-Involvement, Motivation and Demographic Variables on Organizational Commitment: Evidence From it Industry. *International Journal of Business and Management,* 9 (12), 1833-8119.

Thomas, K. W. (1992). Conflict and conflict management: reflections and update. *Journal of Organizations Behavior,* 13 (3), 265-274.

Tjosvold, D. (2008). The conflict-positive organization: it depends upon us. *Journal of Organizational Behavior,* 29, 19-28.

Tourism of Portugal (2007). Site: www.turismodeportugal.pt

Urošević, S. & Milijić, N (2012). *Influence of Demographic Factors on Employee Satisfaction and Motivation.* 45 (4) 174-182.

Wagner & Hollenbeck (2002). *Comportamento Organizacional.* Editora: Saraiva.

Wall, J. A., Blum, M. E. (1991). Negotiation, *Journal of Management,* 17, 273-303.

Warr, P. (2008). Work values: Some demographic and cultural correlates. *Journal of Occupational and Organizational Psychology,* 8, 751-77.

Wegner, Bohnacker, Mempel, Teubel & Schüler (2014). Explicit and implicit affiliation motives predict verbal and nonverbal social behavior in sports competition. *Psychology of Sport and Exercise,* 15, 588-595

Annex I: Request for Authorization to Apply the Questionnaire

Rubina Patrícia Martins de Sousa

Travessa dos Loureiros nº 4 C

9125-112 Caniço

Your Excellency Director

I have a Bachelor's Degree in Management from the University of Madeira and I am currently doing my Master's Degree in Human Resources Management at the University of Minho. The Master's dissertation will be developed during the current academic year and will be supervised by Professor Emília Fernandes.

The research I propose to do has as its theme "Conflict Management and the Motivation of Hospitality Professionals", having as main objective the characterization and resolution of the various types of conflict, as well as understanding the main motivations of hospitality professionals.

Thus, I hereby request your collaboration in my research by completing the questionnaires, ensuring their confidentiality and privacy. The results of the questionnaires will be presented in aggregate form. These will be made available later for possible consultation.

Your contribution is essential to this study and, without it, the main objective of this research cannot be achieved. If you authorise me to carry out the surveys, I will immediately distribute and collect them.

I am certain that this request will be well understood and appreciated, and I am enclosing a copy of the questionnaire.

Thank you for your attention and I am available for any clarification.

Rubina Sousa

Contacts: rubina1991@hotmail.com

Appendix II: Sociodemographic questionnaire

<u>QUESTIONNAIRE</u>: This survey arises within the scope of the Master of Human Resource Management and aims to assist in the collection of data for the dissertation.

<u>INSTRUCTIONS FOR COMPLETION:</u>

- This questionnaire is **anonymous and confidential**. In that sense, it is not necessary to put your name.
- **There are no right or wrong answers in** this questionnaire, it is just your opinion.
- Please be **sure to answer** any questions.
- Once you have completed the questionnaire, **insert it into the envelope and give it to your Manager or Person in Charge.**

<u>PART A:</u> PERSONAL CHARACTERIZATION

- **Please tick or complete the blank space**

1. AGE: _______

2. SEX: F M ☐ ☐

3. DURATION OF PROFESSIONAL ACTIVITY: _______ (years)

4. ACADEMIC QUALIFICATIONS:

   1st Elementary School Cicle ☐

   2nd Cycle of Basic Education ☐

   3rd Cycle of Basic Education ☐

   Secondary Education ☐

   Bachelor's Degree ☐

   Degree ☐

   Master's Degree ☐

   Another ____________________

5. EMPLOYMENT RELATIONSHIP:

   Belongs to the Board ☐

   Contractor ☐

Another ____________________

6. WHICH MOTIVATED HIM TO WORK AT THE HOTEL:

   Family reason ☐

   Interest in the area ☐

   Knowledge or Experience in the area ☐

   Possibility of career development ☐

   Accessibility ☐

   Only one job opportunity ☐

   Other____________________

7. DOES THE HOTEL PROVIDE TRAINING COURSES?
   Yes ☐
   No ☐

8. DO YOU HOLD A MANAGERIAL POSITION IN THE ORGANISATION?
   Yes No ☐ ☐

Appendix III: **Conflict Management Questionnaire**

When you are faced with a conflict situation, how often do you adopt these behaviors? **Mark the** following box **with a cross,** using the following scale from 1 to 5, 1 being "rarely", 3 "sometimes" and 5 "always".

Behaviors	Rarely	Occasionally	Sometimes	Often	Always
	1	2	3	4	5
1. I defend my position with tenacity.					
2. I try to put the needs of others above my own.					
3. I try to reach a compromise acceptable to both parties.					
4. I try not to get involved in conflicts.					
5.I try to examine the problems together in a comprehensive manner.					
6. I try to identify what is wrong with the other person's position.					
7. I seek to promote harmony.					
8. I negotiate to get some of what I propose.					
9. I avoid opening discussions about controversial aspects.					
10. I openly share with others the information I have in order to resolve points of contention.					
11. I like to win an argument.					
12. I go along with the suggestions of others.					
13. I look for a middle ground to resolve disagreements.					
14. I keep what I feel to myself in order to avoid misunderstandings.					
15. I encourage an open exchange about disagreements and problems.					
16. I find it difficult to admit that I am wrong.					
17. I try to help others to avoid "loss of posture".					
18. I emphasize the advantages of "give and take".					
19. I encourage others to take the initiative in resolving the controversy.					
20. I present my position as being only one point of view.					

Appendix IV: **Motivation Questionnaire**

Describe your situation at the workplace. **Mark with a cross** using the following scale from 1 to 7, 1 being "never", 3 "rarely" and 7 "always".

Questions	Never	Almost ever	Rarely	Some times	Usually	Almost always	Always
	1	**2**	**3**	**4**	**5**	**6**	**7**
1. I like to constantly improve my personal skills.							
2. I like to be supportive of other people, even if they are not my relations.							
3. I have a secret desire to get people's attention.							
4. I strive to improve on my previous results.							
5. I feel satisfaction when I see that a person who has asked me for help is happy with my support.							
6. I insist on a certain opinion just to "not give the arm to be twisted".							
7. I like to know whether or not my work was well done so that I can do better in the future.							
8. If I had to dismiss a person, I would try above all to understand his feelings and support him in whatever way I could.							
9. I have arguments with others because I usually insist on what I think should be done.							
10. At work, I try to do better and better.							
11. At work, I like to be a kind person.							
12. I try to relate to influential people.							
13. I try to do my work in an innovative way.							
14. I feel satisfied working with people who like me.							
15. If I can call people into my team's work, I look for those who allow me to exert the most influence.							
16. At work, I pay close attention to the feelings of others.							
17. When I participate in a social gathering, I use the opportunity to influence others and get their support for what I want to do.							
18. I am concerned when I feel that I contribute in any way to the unease of relationships at work.							

THANK YOU VERY MUCH FOR YOUR COOPERATION.

Please, after completing the questionnaire, **check that you have answered all the questions**

I want morebooks!

Buy your books fast and straightforward online - at one of world's fastest growing online book stores! Environmentally sound due to Print-on-Demand technologies.

Buy your books online at
www.morebooks.shop

Kaufen Sie Ihre Bücher schnell und unkompliziert online – auf einer der am schnellsten wachsenden Buchhandelsplattformen weltweit! Dank Print-On-Demand umwelt- und ressourcenschonend produzi ert.

Bücher schneller online kaufen
www.morebooks.shop

KS OmniScriptum Publishing
Brivibas gatve 197
LV-1039 Riga, Latvia
Telefax: +371 686 204 55

info@omniscriptum.com
www.omniscriptum.com

OMNIScriptum